Practical Systems and Procedures Manual

Practical Systems and Procedures Manual

J. Van Duyn

RESTON PUBLISHING COMPANY, INC.
Reston, Virginia
A Prentice-Hall Company

Library of Congress Cataloging in Publication Data

Van Duyn, Julia
Practical systems & procedures manual.

1. System analysis. 2. Electronic digital computers
–Programming. I. Title.
QA402.V36 001.6'1 74-30160
ISBN 0-87909-634-9

A Prentice-Hall Company
Reston, Virginia 22090

10 9 8 7 6 5 4

Printed in the United States of America.

Contents

Illustrations

Preface

The purpose of this book is to provide the reader with a practical and useful up-to-date reference manual on systems and procedures. This handy guide can be used by systems men, procedures analysts, staff personnel making management studies, and students of systems and procedures.

For ease of reference, the book is divided into two parts: Part 1, Systems, and Part 2, Procedures. Part 1 provides a narrative blue print for system development, each section and subsection being illustrated with sample applications. Part 2 describes various types of currently used procedures with sample formats and detailed instructions on how to prepare the procedures.

This handbook may be used by large or small companies, regardless of the type of manual or computer operations they may have. Moreover, the guidelines within this book are valid in any commercial or academic environment.

Finally, for effective communication, the emphasis throughout the book is on concise and legible writing, and easy-to-follow instructions.

J. Van Duyn

Introduction

In the past few years the concept of "systems and procedures" has changed drastically. Systems concept has evolved into a specialized, sophisticated methodology that has become an integral part—indeed the basic structure—of most business, government, research and academic activities. Systems or Systems Development Departments and variations of these titles have generally replaced "Systems and Procedures Departments." Experienced systems men have become sought-after specialists, who concern themselves primarily with analyzing and identifying needs and problems; proposing the most advantageous solutions based on cost/savings and other factors, and designing systems—automated and/or manual—to fulfill the needs and resolve the problems in all types of environments.

Procedures concept, though still a support function of systems, has become in most companies a department in itself. Written procedures are the means through which systems men ensure that new or revised systems will be performed as designed. Knowledgeable procedures men have become procedures and standards analysts who analyze procedures and work flows, identify procedural bottlenecks, recommend solutions, write functional procedures, and develop and set up standard procedures for all staff activities within the organization.

A/ CURRENT ROLE OF SYSTEMS AND PROCEDURES IN THE OVERALL MANAGEMENT JOB

Because today's highly developed systems (especially data processing systems, which are used by an overwhelming majority of companies) are expensive, cost-conscious management employs the latest techniques to optimize systems and its supporting function: Procedures.

Since management is concerned with financial viability, their interest in systems tends to be *what* a system does, and not *how* it is accomplished. In the case of procedures, management is interested in using fewer forms, less paper work and more meaningful, lucid procedures. In sum, management expects—and rightly so—both systems and procedures to help the organization function optimally with minimum cost.

The "how" of systems is entrusted to systems men. And the systems men, by establishing true requirements, eliminating redundancy and identifying constraints that will be imposed on the structure of the systems because of scheduling, technical feasibility, policy and legal factors, fulfill the real role of systems.

Nevertheless, to ensure the compatibility of system projects with the goals of the corporation and the quality of the required information produced, management has become directly involved in system projects. This active participation, which includes keeping a check on the projects and operation costs so that they do not exceed the estimates, is via "steering committee," "policy committee," or "review committee."

These committees plan, review and directly control all system development projects within the company. Further, they establish quantitative criteria as to system projects' benefits, such as cash and resources required, potential return on investment and how the solution provided will affect the long range goals of the company. In addition, they apply qualitative criteria to the projects' possible effect on personnel, equipment and operations, and their ramifications throughout the organization. Finally, the committees consider what the competitors are doing about the same type of problem(s).

As stated previously, the majority of service and manufacturing companies, government, academic institutions and hospitals utilize data processing systems. Nevertheless, there are still many warehouses, factories and shops that use manual systems in assembly, processing and maintenance work. There are also many types of clerical and control work that use manual systems. Consequently, the system development phases summarized below and described in detail in Part 1, apply to both manual and automated systems except, of course, the computer program specifications and the programming phases.

B/ CONCEPTS AND APPROACHES TO PLANNING AND CONTROLLING SYSTEM PROJECTS

The latest approach to system development is the concept of structured series of activities. By defining a structured order of activities, system controls and checkpoints can be established in the Planning Phase, thus ensuring quality and reducing cost.

Briefly, here are the five structured phases:

1/ Planning Phase

a/ *Problem Analysis*

In this initial effort, the systems man gains an overview or a "big picture" of the organization. He gathers the symptoms of the problem; plans carefully the subject and scope of the preliminary study, and writes up the assignment. At the end of this step, he secures a sign-off approval on the written assignment. This insures the systems analyst against any future misunderstanding or arguments.

b/ *Interviews*

The systems man plans whom to interview and what specifics to ask. Subsequently, he interviews key people, using an openminded, positive approach to get the maximum benefits out of the interviews. He also draws a chart to show the current flow of work. He carefully studies all of the data and then plans and writes a proposal.

c/ *Proposal*

Within the proposal he states the need and/or problem briefly. He defines the scope and nature of the problem(s), recommends the best solution and points up the possible benefits.

2/ Development Phase

a/ *System Requirements*

The systems man establishes the system requirements by performing a comprehensive systems study, i.e., by analyzing the procedures, forms, job functions, inputs, outputs, reports, files and controls of the present system (if there is one). As a final step of this task, he writes a brief present system narrative.

b/ *General System Design*

The systems man designs the new system, draws the general system flowchart and writes a general system narrative, so that the user can get a visual and narrative overview of how the new system will work.

c/ *Checkpoint/Sign-Off*

At this milestone, the documented data from the preceding step are given to the user. Subsequently, both the user and you sign off on the general system design. This, in effect, constitutes a contract ensuring that the user will not add any other requirement; that the systems man will deliver the system as stated in the above documentation, and as scheduled.

3/ Solution Phase

a/ *Detailed System Design*

The systems man designs the new system in detail.

b/ *Program Specifications*

The systems man establishes the program specifications, thus providing all the needed and required data to the programmers.

c/ *Programming*

The programmers—under the systems man's supervision—code, test and debug the programs.

d/ *System Testing*

The systems man tests the system to ensure that it works as he designed it.

4/ Installation Phase

a/ *Plan Installation*

The systems man plans carefully the how, what and where of the installation of the new system.

b/ *User Training and User Guide Manual*

The systems man ensures that the user is trained to use the new system. He also provides the user with a functional guide manual.

c/ *Installation/Conversion*

The systems man installs the new system and, if any conversion is to be done, supervises that activity also.

5/ Post-Installation Phase

a/ *Auditing/Evaluating*

In this, the last phase of system development, the systems analyst gives the documented new or modified system to the auditor for auditing and evaluation.

b/ *Final Report*

The systems man writes a report to the user/management on the progress and results of the installed system. The results should reflect careful consideration of the user's requirements, current system techniques, latest data processing methodology, optimum use of available resources and an appropriate degree of system flexibility for modifications or possible future expansion.

C/ DATA BASE MANAGEMENT SYSTEMS

When management in an electronic data processing environment decides that time has come to change the original computer systems and consolidate the many small and often redundant systems (e.g., accounting, engineering, marketing, payroll, production systems) into a Management Information System (MIS), i.e., a general purpose or Generalized Data Base Management System (GDBMS), the task of establishing such system is given to the systems department.

Because the concept of Data Base Management Systems and that of the data base administrator are part of today's computer field, an overview of these topics is a must in an up-to-date book on systems and procedures.

What is a GDBMS? Simply put, it is a funnel of information. For example, in a large bank tens of thousands of checking and savings, commercial and personal loans, personal and investment trusts, and other data accounts are input into a GDMBS. Subsequently, on request, the needed data are compiled, manipulated and verified, and reports or statements are produced. The reports can range from budget, budget forecasting and profitability, to cost accounting, branch site selection, market penetration, and so forth. GDBMS can also be likened to a warehouse that receives and stores all given information in one place, therefore saving time, space and effort to get the data from many places. The main concept of such systems is to eliminate many small systems and consolidate into one integrated system.

A well-designed GDBMS should offer the following advantages to the user:

- A general inquiry capability to make the system a useful and accessible management tool.
- Ability to permit authorized users to share common information.
- A means for nonprogrammers to input data, manipulate data files, retrieve records and ask for and get formatted reports, using English-like language.
- Cost reduction via centralized control of data and system maintenance.
- Functional flexibility to tie together all the data files so that users can work with them as a single integrated data base.
- Capability to handle both fixed- and variable-length information records to use the available storage most efficiently.
- Rapid system response to changing requirements.

Once such a centralized system is established, the data base administrator's role is to ensure that it is fully utilized and maintained. Further, he is responsible to maintain the security of the data base. Specifically, he directs the development of sound methods for protection against possible destruction or access by unauthorized people and recovery from possible disaster. His function is to keep users informed of changes and conditions affecting the system, and provide them technical assistance. In short, the data base administrator's work is highly sensitive and complex and vitally important to the entire organization.

D/ SYSTEMS DEPARTMENT

A systems department's size, structure and organization depends on the size and type of organization within which it is established. Generally, however, it is headed by a manager who might report to the director of the data processing center if the systems department is in a computer environment, to the controller or to a vice president. The systems department manager is usually on the same level with, for example, the engineering department manager, the marketing department manager or operations department manager.

At most companies the structure of the systems department is such that under the manager the next level is that of section supervisors. In turn, project leaders or senior systems analysts head project teams and report to the supervisors. These project leaders have to be knowledgeable in both systems analysis and computer programming, and of the specific field the company is operating in, since they are responsible for the success or failure of their systems projects. The project leaders directly supervise systems analysts and computer programmers, and sometimes keypunch operators.

The systems department manager and his staff are responsible for planning, developing, implementing, and maintaining new or revised systems, manual or computer, along with established systems within the firm.

Regardless of the class or dimension of company within which it is established, the systems department is a service organization that responds to various departments' request within the company. Systems men, depending on what field of endeavor they happen to be in, are given problems such as: to design a system that will expedite the processing time of loan applications in the installment loan section of a department store; to design a system that will streamline a bank's stock transfer processing; to design an effective inventory control system in an electronic parts manufacturing company; to design an efficient document control system in an engineering organization.

1/ Developing and Establishing Standards

The development and establishment of standards within the systems department is imperative if there is to be uniform efficiency in the designed systems. What is "standards"? Standards is a stringently defined means by which certain work is being performed, whether it is shipping operations, data processing, assembly work or the numbering the company is using in all of its documentation. It is the responsibility of management to develop, establish, and enforce standards throughout the company. Further, when certain conditions require exceptions to the established standards, the extraordinary procedures must have the approval of the management. The responsibility of ensuring that all staff adheres to the established standards starts with the systems department manager and continues to the section supervisors and the project leaders.

Clearly defined and up-to-date standards are needed for job names; for manual and computer operations; for input media and output media; for control procedures; for name/address for all systems; for installation/conversion; for system testing; and for documentation. In addition, there is a need for standards for interviewing job applicants; for personal evaluation; for promotion and salary increases; and for obtaining outside consultants, contract programmers, draftsmen and so forth. However, it cannot be emphasized too much that only when management is committed to the enforcement of standards will the benefits of standardization be realized.

2/ Qualifications of Systems Men

A prerequisite for a systems man or woman is a keen sense of "problem solving" and the ability to communicate effectively with all levels of management and staff. Whether he is in manufacturing, insurance, finance (the fields that, according to statistics, employ the majority of systems analysts) or any other field, the three basic phases of his work are:

- Gathering data and studying carefully both symptoms and problems or needs.
- Analyzing and evaluating all data and, based upon the findings and management-established quantitative and qualitative criteria, recommending the best method of solution.
- Designing a new system or revising the old system and implementing it.

Within these stages the systems man plans and executes interviews with user management and personnel to obtain information requirements at each level of activity that the present system supports. He must remember that he is gathering data about problems and flaws in systems and *not* in people. He must identify constraints within which the system will have to be structured; determine the controls that will be necessary to ensure the validity of information processed or produced by the new system; formalize system output representation; and ascertain the benefits that may be expected from the new system. In addition to being thoroughly familiar with the type of company within which he works, he must do comprehensive ground work if he is to design and implement an efficient new system.

It has been estimated that by 1975 the number of computers in the United States alone will reach some 150,000, thus it is probable that most systems analysts are or will be working for companies in which computer systems are intrinsic parts of the organizations. Data processing environments in turn necessitate formal education and training.

While an undergraduate degree or at least two years of college is a definite requirement, it can be in almost any discipline. However, if the systems man majors in computer science and minors in business administration he will have no trouble in securing a junior systems analyst position at any company. In addition to theoretical knowledge of computer languages and computer systems such as IBM, CDC, Burroughs and PDP, the systems analyst must be proficient in machine language programming and high-level programming languages such as COBOL and PLI. Because first hand knowledge of computer programming is so important in systems analysis, computer programmers are often promoted first to junior or assistant systems analysts and then to systems analysts and senior systems analysts. However, it is usually programmers who are willing to take evening courses in systems analysis and allied disciplines at local colleges and adult education who will be given opportunities to advance. It is also advisable for future systems men and women to attend seminars and workshops given by professional associations such as Association for Systems Management, American Management Association and American Computer Machinery Association (ACM).

3/ Procedures and Manuals Department

This department or section might be called Procedures and Manuals, Procedures and Standards or Technical Publications. However, whatever the title, it supports the systems department. As such, it has to adhere strictly in all its end products—procedures, manuals, and so forth—to the established standards. These standards are set up to indicate *what* must be done, while procedures describe in a concise, legible manner *how* to perform a specific task.

Generally, the procedures department is headed by a manager who might report to the manager of the systems department, or to a vice president, or to the director of the data processing center (if the firm is large enough to have one). Under the manager of the procedures department usually there are the supervisors of various groups such as: manuals group, control procedures group, operations procedures group, development manuals group. The procedures department, just as the systems department, is a service organization. It fulfills the need for written instructions in all phases of operations, processing, development and policy of the company. Consequently, the parameters applied to the activities of the systems department apply equally to the procedures department.

4/ Qualifications of Procedures Analysts

The requirements of procedures analysts are somewhat similar to that of systems analysts. He or she also must have the ability to communicate effectively with various levels of management and staff. The procedures analyst, however, in addition to communicating verbally has to communicate via the written word as well. The procedures analyst must remember at all times that the most important word in his vocabulary is "why?" and that his most important skill is strict attention to detail.

Whether preparing a manual or computer processing procedures, the three basic phases of his work are:

- Gathering data about present procedures, i.e., how a certain work is processed, and studying carefully the problems and flaws in the flow of work.
- Analyzing, evaluating and organizing all data and, based upon the findings and the management-established con-

straints, recommending the best method for solving the problem or the performance of a particular section.

- Establishing standards through newly designed forms. Documenting in concise, legible, easy-to-follow procedures.

Within these stages the procedures/standards analyst must carefully plan and execute interviews with user management and personnel to obtain a true picture of current processing and procedures. When interviewing or observing the performance of a task, he must remember never to be critical of either the employee being interviewed or observed, or of anyone else in the organization. He must draw a work or grid flowchart of the present flow of work in each department, and also as the work is channeled from one department to another, until the cycle of processing is completed.

The procedures analyst must be thoroughly familiar with the type of company within which he is employed. The ground work that he does to provide the organization with functional procedures must be just as thorough as that of the systems man. Whether he is in manual or data processing environment, the procedures analyst must have an undergraduate degree or at least two years of college. Preferably, his major should be in English, Language Arts or Journalism because good grammar is a prerequisite and a large vocabulary is highly desirable. Indeed, the procedures analyst must understand that he is a "communicator," and he or she must be guided by the criteria: Does the finished product—procedures, instructions, flowcharts, manuals, studies—really communicate the essential thought, the principal philosophy of the designer of the system, and does it conform to the company established standards?

5/ Outside Consultants and Packaged Services

In the past few years the number and quality of outside consultants and packaged services have grown into a successful industry along with the burgeoning electronic data processing field.

It is not uncommon for management to call in outside consultants. These consultants are often contracted to provide computer programmers for a high priority, limited time project, revise the flow of work in a certain department, establish and document standards in the systems development section of the systems department and design and implement data management systems. Consultants are called in sometimes to completely reorganize the systems department or some other department that shows slow productivity, unusually high personnel turnover or other activities that displease management. Further, consultants are called in when the company needs specialists to teach/train new or inexperienced personnel in such areas as systems analysis, programming languages, ANSI (American National Standard Institute) flowcharting methods or effective technical/business writing.

Before signing a contract with an outside consulting firm, a decision-making process must be established by management if the service package is to be successful and have lasting results in the particular area and on its personnel.

Management should start this important step by a) establishing and documenting the need or the problem, b) setting up the selection criteria, c) searching and screening all available outside consultants who are known for their expertise in the specific area, d) evaluating the consultants by comparison of documented sample products and price and e) evaluating cost. In addition, management should seek staff support by informing them of the decision to use outside consultants. Above all, management must be methodical in their approach to evaluate and select the outside consultants if they want the most for their money in terms of a lasting solution to some problem or need in the organization.

The minimum criteria by which consulting firms should be evaluated and selected are:

- Length of time in business.
- Dun & Bradstreet rating.
- List of clients.
- References by satisfied clients.
- Number of staff and their qualifications.
- Quality of documentation/proposal.

The last but certainly not the least factor in evaluating the services of outside consultants is the big question, "What is the package price of the services offered?"

There are many advantages for using outside consultants in special or sensitive projects and areas. One of these advantages is the known impartiality of the consultants. This is especially important in the case of improving work flow in a certain section of an organization, or establishing standards and procedures in another section or department. The staff will accept even unpopular recommendations by outside consultants much more readily than by company analysts, regardless of how competent the latter may be.

Part 1

Systems

To be efficient and functional and to meet the requirements of the user/management, an automated or manual system must be premised on the following basics: It must be tailor made to the particular user and constrained by qualitative and quantitative factors peculiar to the specific business or environment.

In addition to the above criteria, the system development project should be set up within a task-oriented framework, and the matter of cost and time should be approached with extreme caution. The department head or manager who sends in the service request is usually impatient to have the problem solved; the job done. Consequently, he might try to pressure the analyst too early in the project into a commitment on cost and time to-be-extended. Quite often the manager of the particular department has only a hazy idea of what is involved in systems planning and development. Therefore, the systems man must point out that to design and develop a system that makes operations more efficient, services more useful and management more effective, the analyst must invest sufficient time on the preliminary or planning phase before any estimates on cost and time span can be given. In other words, gaining background knowledge, analyzing the service request, investigating the symptoms and the problems, defining the requirements and working out all possible alternatives are necessary tasks before either time or cost can be estimated.

Further, it is imperative that a comprehensive feasibility study be conducted in the development phase so that the analyst can give definitive answers to the questions: Is the proposed system operationally feasible? Technically feasible? Economically feasible?

The total effect of the system development efforts should be an integrated planned approach to meet the automated and/or manual processing needs of the enterprise. Thus, no system development should be a stand-alone, unrelated job. It should interface with other systems (if there are other systems) and be compatible with the goals of the organization.

To be in tune with corporate goals, the analyst—before starting the system development project—should have a clear understanding of the overall objectives of both the corporation and the requestor's department. He should study the communications (or lack of it) within the department and with other departments; review to some extent management and personnel relationships; check the department's relationship with user/customer and investigate its potential growth. This last variable is very important insofar as the design of the system is concerned, because even though any system that is well designed is flexible/modular and "open-ended," the growth factor is crucial when designing a system.

Since the decision by management to "buy" or "kill" the project is based on the submitted proposal, it is an absolute necessity that the analyst perform a comprehensive basic investigation so that he is thoroughly familiar with not only the problems but also the business and environmental factors before he starts writing the proposal.

In addition, the proposal should be written in language the management can easily understand. This means a minimum of computer/technical jargon, and no computer printouts or programs. Moreover, he should describe *what* the new or modified system will do, and not detail *how* the system will work. In general, management feel that the "How" of the system is the responsibility of systems people.

As important as the proposal is, legible, up-to-date documentation of every phase, every activity within System Development is just as important. In fact, by establishing the standard that an activity is NOT finished, and that the next activity CANNOT be started until that particular activity's documentation is complete, the systems analyst ensures that an up-to-date, effective documentation is performed. Such documentation is also a good systems control tool.

In sum, the analyst should get the big picture, the organizational overview first, and subsequently, in well-defined steps, perform and document the five structured system phases detailed and illustrated with sample applications and standard forms in the following chapters for a cost-and-time controlled system development project.

Chapter
1

Planning Phase

This chapter instructs in detail the technique to be followed in identifying each task within the Planning Phase to effect a structured system development.

Planning phase is perhaps the most important activity in system development since it is the foundation of any system. Most system failures can be traced to a careless or superficial planning phase.

It is during this phase that apparent and actual problems are identified; problems' scope and urgency defined; alternatives for possible solutions examined; the effect of the proposed new or modified system on personnel, equipment and corporate goals considered and "ball-park" figures of cost and time estimated. A properly executed planning phase will ensure that systems controls and checkpoints are established that assure a quality product and reduce the cost of the system development project.

The planning phase consists of three tasks: A/ problem analysis, B/ interviews and C/ proposal.

Within these three tasks there are two checkpoints: The first one is at the end of problem analysis, when the analyst writes up the assignment after completing the preliminary study of the organization and the problem areas. This document, which restates and redefines the service request, is presented to the requestor for approval-signature, and it serves as an insurance against future misunderstanding.

The second checkpoint occurs when the proposal is presented to management, and the proposed system is either accepted or rejected.

When gathering the symptoms in this phase, and later when doing the system study in the development phase, the analyst should look at the problems and the possible solutions through the eyes of the management/user, and not through the eyes of a systems man or programmer.

The project's feasibility is evaluated in terms of development, implementation, conversion, operating costs and potential benefits based on return-on-investment. These estimates of costs and benefits should be based on an in-depth service request analysis, interviews with management and staff to determine the effect of the new system on personnel and equipment, and on the preliminary design of the proposed system. This approach works best when it involves close interfacing with the user, and the user is heavily committed in the proposed system.

Certain steps in the problem analysis, such as gaining an overview via organization charts, written office communications and memos, as well as planning the preliminary study of whom to see and what specifics to ask, should take minimum time. Other steps, such as analyzing service request and investigating flow of work and processing of data, should be comprehensive, and the time spent should be restricted only by the matrix of the whole project. An in-depth, thorough service request analysis shortens the following phases in system development and makes those tasks much easier.

For the interview activity the analyst should plan carefully, prepare a list of specific questions and never leave anything to memory. He should take notes or, better yet, if the interviewee does not object, record the interview on a cassette tape recorder. Finally, the analyst should never give the impression that he is in a hurry, that he is impatient to finish the interview. Such attitude is poor, and its effect is an unresponsive interviewee.

Preparing, writing and submitting the proposal requires the analyst to organize and sift the material he gathered, write clearly, concisely and cohesively and "package" it attractively.

Some systems men offer many alternatives in their proposals, leaving the decision up to management. Such approach is negative, and tends to bring negative reaction from management. Management feels, and rightly so, that *you* are the systems analyst, *you* should evaluate all possible alternatives and *you* should select and offer management the best solution based on your research and analysis of all data.

At the start of the planning phase, if you will establish "incremental documentation," and all through the system development keep it within one week of technical reality, it will pay high dividends.

Incremental documentation is a technique of having questions (which subsequently are replaced by answers), detailed descriptions, interviews, charts and schematics typed/drawn immediately and kept up-to-date in a documentation work book. This type of documentation is invaluable as a communication tool since each project member as well as the user is provided with a photocopy of the continuously updated work book during system development. Thus, incremental documentation serves as a major aid in monitoring the project and, in addition, it greatly simplifies and shortens the task of having the standard documentation complete at the end of each phase before the next phase can be started.

A/ PROBLEM ANALYSIS

The purpose of the problem analysis task is to establish a solid foundation for problem definition and system requirement by identifying the requestor and his present and future needs and problems. This initial task of the planning phase consists of four distinctive steps: 1/ gaining an overview or the "big picture" of the organization and of the project, 2/ analyzing and developing basic information about the service request by gathering symptoms of the problem, 3/ planning carefully the preliminary study and 4/ writing up the assignment.

1/ Getting the "Big Picture"

To set up the system development project so as to support the corporate goals, i.e., to be in tune with the corporation, you must have a clear understanding of the organization, of the long range plans of both the enterprise and the requestor's department and of the latter's growth. You should study organization charts (Figure 1-1) and read office correspondence and memos to determine the level of communication within the department in which the problem exists. In addition, you should investigate the department's relationship with other departments and users/customers, and the relationship of the department management and staff. A clear overview will help you to evaluate in proper perspective the service request and the symptoms of the problem.

All through this and other activities you should make copious notes that should be typed daily and placed in your documentation work book.

2/ Analyzing the Service Request

Before gathering symptoms of the problem, study and analyze the service request thoroughly, so that at the end of this step you are able to define:

- Who is the requestor?
- Who are the customers of the requestor?
- What type of services does the requestor need now?
- At what frequencies are these services needed?
- What type of services will the requestor need in the future?

It cannot be repeated often enough that you have to do in-depth research when analyzing the service request. Thus, when you find that the requestor is management, further defining is necessary since management comprises many types and levels. The services that the sales or marketing manager needs are quite different from those that the engineering manager or purchasing manager need. Service request analysis begins by identifying the requestor; his present and future needs and problems. It is

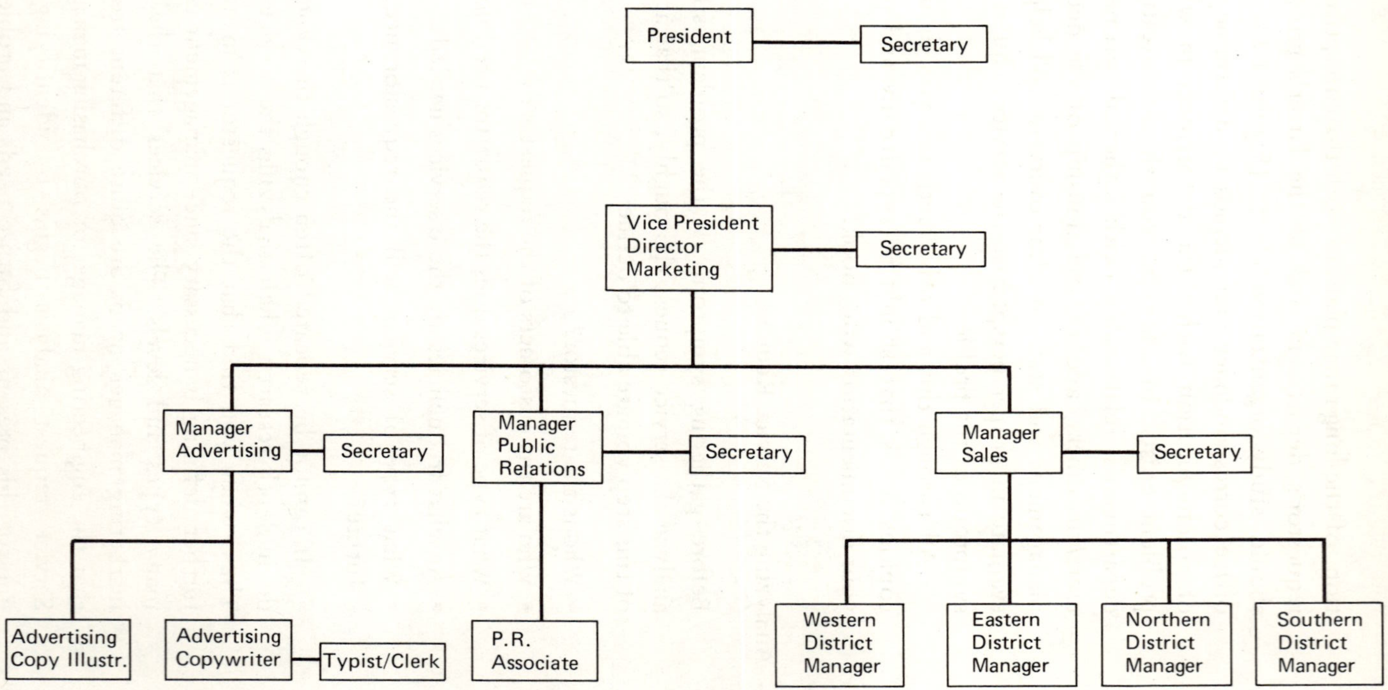

Fig. 1-1 Organization Chart

very important, however, that all through this activity you keep in mind that the requestor is describing the symptoms and not the problem. It is up to you through investigation and analysis to ascertain the actual problem. You can start searching for the real problem by reading department documents and communications, and writing down your observations, questions and comments in response to the various communiques and the descriptions of various procedures. You should draw a work flowchart (Figure 1-2) using standard flowchart symbols (Figure 1-3) as you observe at first hand the work of the involved personnel. Subsequently, by studying all the notes you have made about the symptoms, and relating them to your work flowchart of the existing flow of processing, you can begin to identify the problem areas.

When you believe that you have recognized the problem areas, tentatively define the urgency and scope of the problems. At this stage, however, do not attempt to arrive at alternatives for resolving the problems. You do not have enough data for that. You will have a much better idea of the problems and the possible alternatives to resolve the problems after you have laid out the plans for your fact-finding preliminary study and have conducted your carefully prepared interviews with management and key personnel.

3/ Planning the Preliminary Study

The more detailed your work is in this task, the more vital information you will get to help you correctly diagnose the problems at a later phase. Studying your flowchart of how the work is being processed in the requestor's department will point out to you the blank spots in your information gathering. The next logical step is to prepare a list of questions. Through the response to these questions later in the interview, you will find out what management and the requestor know, and what they need to know–i.e.,

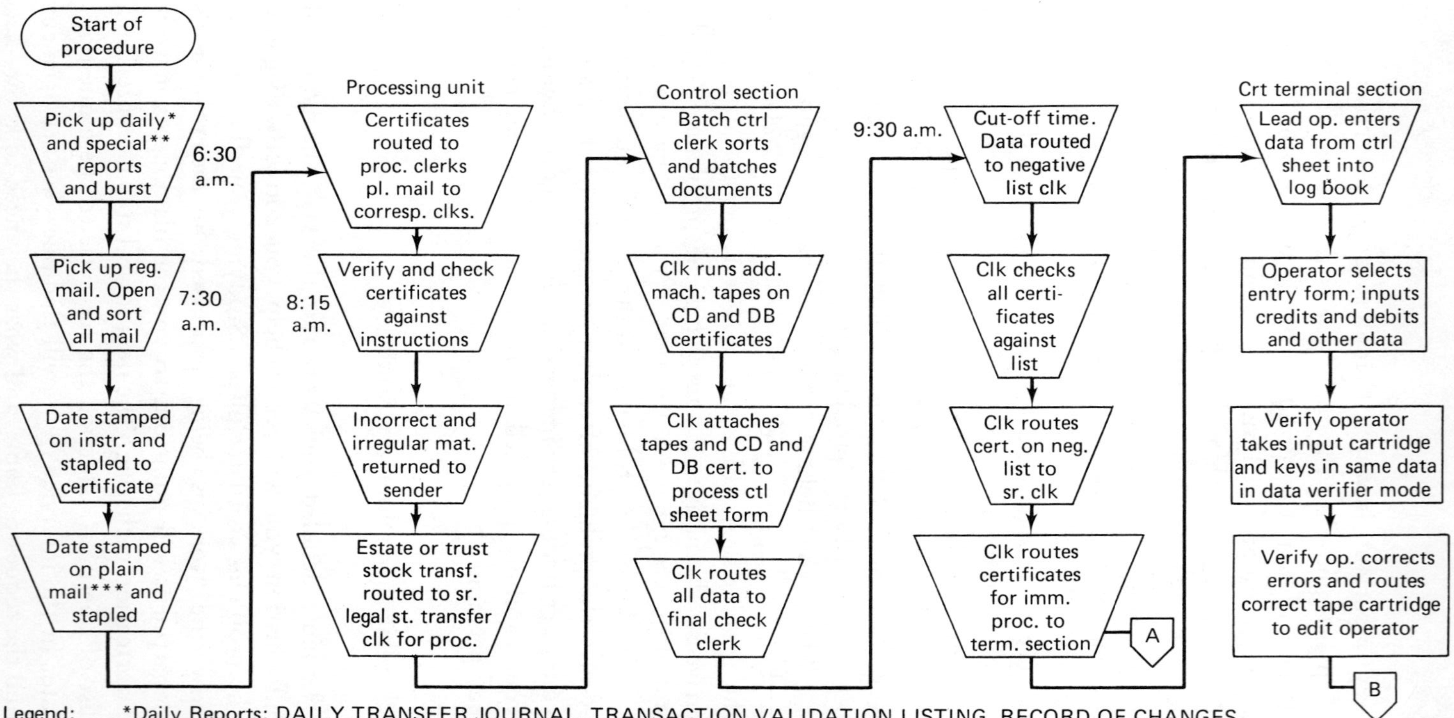

Legend: *Daily Reports: DAILY TRANSFER JOURNAL, TRANSACTION VALIDATION LISTING, RECORD OF CHANGES
**Special Reports: SHAREHOLDERS LISTING, DIVISION AND PROXY EXTRACTIONS, ACCOUNT ANALYSIS
***Plain Mail: Correspondence re.; change of address, name, or soc. sec. no., stop-payment notice, etc.

Fig. 1-2 Stock Transfer Work Flowchart

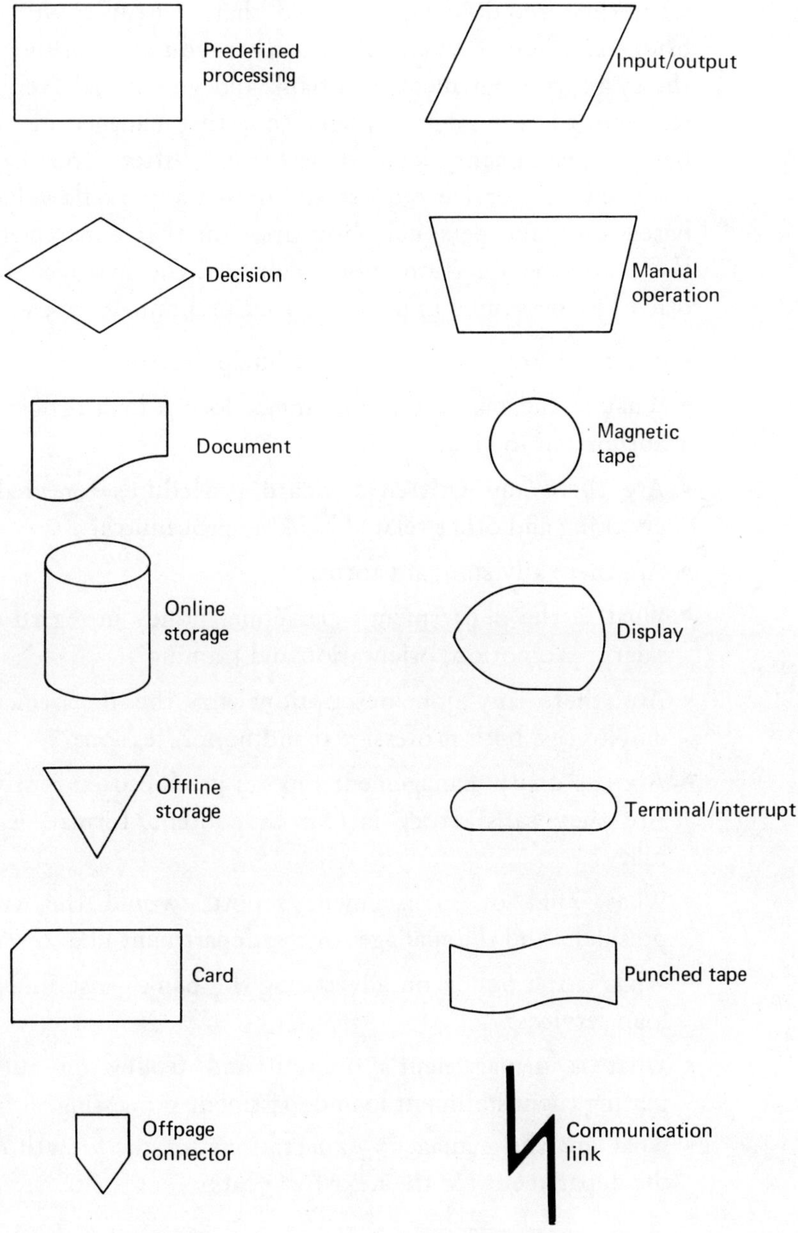

Fig. 1-3 Standard Flowchart Symbols

what their requirements are—and then it is up to you to bridge the gap. For example, assume you are working in the systems department of a bank and you are selected to respond to a service request from the manager of the bank's installment loan department. After you have analyzed the service request and drawn a work flowchart based on your personal observation in that department, these are the questions you might ask the manager and other key personnel in planning your preliminary study.

- Are there any written bank lending policies?
- What is the maximum amount of loan a loan officer is authorized to give?
- Are there any written standard procedures for credit checking and other related banking procedures?
- Are there any standard forms?
- What is the department's personnel policy in regard to salary, promotion, orientation and training?
- Are there any job descriptions for the department employees, both professional and nonprofessional?
- Are there any management reports produced currently? Are they satisfactory insofar as content, format, legibility?
- What kind of management reports would the vice president and the manager of the department like to see?
- What is the policy on advertising the bank's installment loan services?
- What is management's opinion and feeling on automating the installment loan department processing?
- What are the manager's expectations for the growth of the department for the next five years?

4/ Writing up the Assignment

A written assignment (Figure 1-4) is a brief, direct summary of the work that the systems analyst will perform. It defines the problem area, but it does not include technical details of why or how the work will be performed. It states clearly that the definition of the problem area, its urgency and scope has been based only on analysis of the service request and initial investigation.

In other words, the written assignment restates what is in the service request, adding only your impression of the urgency and scope of the problem area.

The written assignment serves the following purposes:

- It provides a checkpoint for the requestor/user and you, the systems analyst.
- It ensures a continuous involvement and communication between the requestor and you.
- It commits the requestor when he approves and signs off on the assignment; it commits you to perform the work as specified.
- It insures you against any possible future arguments about your misunderstanding the service request, and about the requestor not being aware of what you intend to do.

You must check your written assignment carefully to ensure that it specifies only those tasks that are in the service request and the tasks that you plan to write about in your proposal.

B/ INTERVIEWS

Interviews are very important to the information gathering function. Interviewing is an art, and a properly conducted interview will provide the most information for the least effort. If you observe the following points when doing this task, you will find that it makes for an easier and more pleasant time for both you and the interviewee.

- Set up the interview for a time that will be the most convenient for the person to be interviewed.

ASSIGNMENT

In accordance with Service Request No. 495, the Systems Group of the Boulder Oil Company's Systems Department will perform the following work according to the schedule presented in Figure 1-5.

(1) Provide automatic credit card verification by putting individual service stations "on-line" with the Company's central computer in San Francisco.

(2) Identify stolen/lost or invalid credit cards via the same system.

(3) Develop a system that is compatible with currently issued credit cards.

(4) Design a system that is simple in operation in order to require minimal training to operate it.

These objectives will be accomplished by the following efforts:

- Design and develop a system that permits automatic credit verification at the point of sale and is also compatible with currently issued credit cards.
- Select a reliable terminal that is reasonable in price and simple to operate, for minimal training of the service station attendants.

(5) Deliver to the Control Department the equipment that will be installed at the service stations, together with a copy of the User Training Manual

______________________	______________________
Systems Department	Control Department

Date: ______________________

Fig. 1-4 Assignment Form

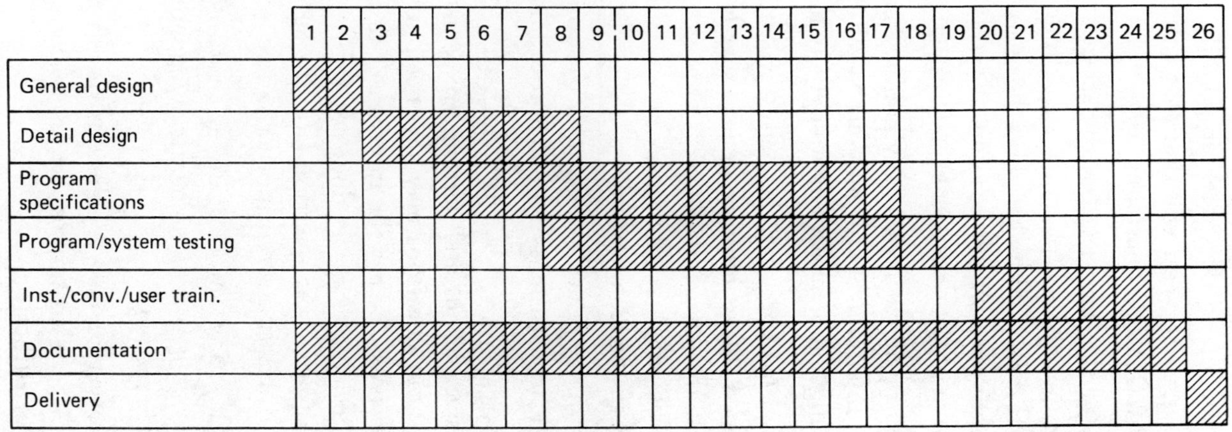

Fig. 1-5 Automatic Credit Card Verification System Activity Chart

- If possible, try to arrange it so that the interview takes place in the office or work space temporarily assigned to you during this project. You will find that this arrangement will keep interruption to a minimum.
- As a basis for the interview, use the list of questions that you have made during the "planning of preliminary study" step. These questions should be based on the analysis of the Service Request and your own observation of the flow of work, plus your studying of the department memos and so forth.
- Try to clear up all vague and hazy points. And remember that the most important word in your vocabulary is "why?," without being argumentative or annoyingly insistent.
- Be friendly. Assume that the person you're interviewing is well informed. Make him/her believe that his cooperation is appreciated, and his contribution is of value. *Never be critical* of either the employee whom you are interviewing or of anyone else in the organization.
- Remember that you are investigating the possible weaknesses and flaws in the existing system and procedure, and not in the employees.
- If the answers you are getting are not informative or conclusive, your questions may have been misunderstood. Of course, it is possible that the interviewee does not want you to know the answer, or he does not know the answer. You can either rephrase the question later in the interview or try to get the information from another source.
- Do not relay any information given to you by another interviewee, even if it is diametrically opposite to what the interviewee is stating.
- Don't indicate that any conclusions of any kind have been reached. Don't try to impress the interviewee with your own knowledge or authority.
- Don't promise a "blue sky" future system. Don't commit yourself to any specific feature in relation to the interviewee's particular work.

- Conclude the interview as soon as the interviewee provides you with the necessary information, and thank him for his time and cooperation.
- Prior to the interview, when you set up the appointment, ask the interviewee if he/she would object to having the interview taped. Point out that by having the interview recorded, there is no chance of his/her being misunderstood. If the person agrees, bring a small cassette recorder. It makes for a much more relaxed interview, since you can give all your attention to the person and not distract him with your writing of his comments.
- If you have to take notes during the interview, keep them to a minimum. And immediately after the interview transcribe and expand your notes into an orderly, legible form. Whether a typist transcribes the tape or your handwritten notes of the interview, make sure that the report bears the name of the employee, your name and the date of the interview.
- All the information you receive in the interviews should be treated with skepticism until the data have been checked and double checked, i.e, verified and confirmed.

C/ PROPOSAL

The purpose of the proposal is to briefly state the need and/or problem, and to recommend the best solution based on in-depth research and analysis of all related data, including consideration of business, environmental and human factors. In addition, the proposal is a selling document that displays your ability to resolve a particular problem and/or need; to undertake the project and to bring the project—via the proposed new or modified system—to successful conclusion. The proposed system may involve only reassignment of responsibilities or redirection of work flow in a particular area. Or, it may call for the introduction of new procedures, new forms and even new policy within a specific department or division. Or, it may suggest the automation of a manual system or the replacement/

modification of an existing computer system. Whatever the recommended system is, the proposal provides management a visible product of your work. Consequently, it should be neatly "packaged," i.e., it should adhere to the standard format used in most published business, technical and academic proposals, and fully respond to the service request by providing the information management needs to make a decision to "buy" or "kill" the proposed system. While the in-house proposal you present to management does not have to be as formal nor as extensive as the document a vendor submits (vendors' proposals contain sections on background (qualifications of the vendor company), related experience, biographies of staff and so forth), your proposal should be logically organized and include the following essential sections: front matter, introduction, discussion, technical approach (only if your proposed system is an automated system), cost analysis and appendix or appendixes if needed.

The front matter consists only of title page, table of contents and list of illustrations and is self explanatory. (For an example of standard front matter see sample application following this section.)

1/ Introduction

The introduction is one of the most important sections of the proposal. It can make or break your proposal since it provides management the initial impression of your capabilities and the soundness of your concepts. Therefore, if you want to "sell" your proposed system take sufficient time to write the introduction concisely, legibly and cohesively. Do not clutter it with details; details belong in the discussion and the technical approach sections of your proposal.

The narrative introduction contains as many of the following subsections as applicable: background information, overview of the apparent and actual problem areas, major concepts of the recommended system, indications of what the proposed system will do and benefits. That is, how the proposed system will benefit management/user, both in tangible and intangible returns. If appropriate, the growth potential and possible limitations of the proposed system may be mentioned.

2/ Discussion

A positive approach and tone should be taken in the discussion section (as well as all through the proposal). Nonetheless, you should not gloss over or ignore any difficult aspects or potential problems. You should present a difficulty or problem as it is, stating, however, that you are confident you will be able to ultimately solve it.

The discussion section comprises the following subsections: narrative description of the proposed system, a grid work flowchart of the present system (or a data flowchart, if it is a computer system), a grid work flowchart of the new or modified system (or a data flowchart, if it is a computer system) and a chart of the schedule of development, estimated benefits and growth potential. Write the narrative description of the proposed system as succintly as possible, but without sacrificing essential information. If you have a mound of supporting data in addition to the few charts or exhibits that illustrate the Discussion for a better comprehension of your recommended system, put your data into one or more appendixes at the end of the proposal. This arrangement will not interrupt the flow of your presentation, yet the data are there for reference for the technical type user.

One of the most effective tools a system analyst has is a work flowchart, which is a detailed, graphic illustration of how a particular type of work is processed from beginning to completion. Sometimes the work flowchart is confined to one department, but more often the processing flows through many departments within the company.

The work flowchart identifies the department(s), section(s), the personnel who process the work and—where the work time is dependent—the time when each function is processed. There are no shortcuts in drawing an accurate, step-by-step work flowchart. In fact, there is only one way of getting this information: follow the processing of a particular work—whether it is customer sales order, production control, loan application procedure, credit card verification or stock transfer procedure—from the very first step to the last step, writing down every step each person does in performing his/her chores.

For example, if you need to study and analyze how a customer sales order is processed in the production department of a MOS integrated circuit company, you have to follow the stack of customer sales orders (CSO) to the production control clerk's desk as he makes up a delivery schedule card (DSC) for each item, and then routes the whole package to the production manager, and so on until the merchandise is shipped, the charges are posted in the individual customer's account and the inventory control is updated. Or, if your assignment is to study and analyze a bank's stock transfer's daily processing, you have to be with the mail clerk at 7:30 a.m. as he receives and opens the mail and stamps the date of arrival on the envelopes and sorts the mail and routes the incoming stocks and bonds to the unit clerks, and the correspondence to the correspondence clerks.

In other words, you have to jot down in your notebook every function of each employee starting with the mail clerk through the unit clerk, control clerk, coding clerk, key-processing clerk and so forth, until the daily processing is complete and the mail clerk sends off the transferred newly issued stocks or bonds. Then study your notes, and draw the work flowchart accordingly (Figure 1-2) using standard flowchart symbols (Figure 1-3).

Several important points should be kept in mind if the chart you are drawing is to be functional and legible to management. They are:

- The work flowchart should be organized according to the frequencies the function are performed. For example the daily processing of a certain function should be on one flowchart; the weekly or monthly processing should be on another flowchart.
- On the first page of the work flow chart a legend or footnotes should explain any abbreviations or acronyms that are used in the flowchart. Also, if, for instance, the daily, monthly or special reports are mentioned in the flowchart, there must be a listing by name of those reports at the bottom of the flowchart. It is frustrating to try to read a work flowchart full of acronyms and statements without any explanation.
- Use a pen and template for drawing the chart, and if possible, typewrite the text within each symbol. A clean, neat, legible work flowchart is appreciated by management and other systems analysts too. Also, it makes the following tasks much easier for you.

3/ Technical Approach

The technical approach includes specific details of the hardware and software components of the proposed system, including new procedures, new forms, and peripheral equipment needed to implement your proposed system. This section also contains the scope of work, program schedule and personnel and project organization.

Finally, the technical approach should include recommendations for manpower control, financial control and time control via program evaluation and review technique (PERT), *if* the proposed system development is large and complex. PERT may be used for all schedules, cost and manpower control, including updating and revising schedules and costs, weekly or monthly as required.

4/ Cost Analysis

The cost analysis is based on return-on-investment (ROI) technique. This methodology involves measurements analysis for specific application(s) area(s). ROI is usually calculated by dividing savings and increased income by development costs, amortized over five years. The development costs are the cost of mainframe and peripheral equipment dedicated to the proposed system, the cost of system design, programming costs and, of course, the overhead. To arrive at an accurate ROI figure, limit the potential profit to savings and extra income that can actually be quantified. To arrive at sound economics do not count any intangible factors such as more effective management, better customer service or more successful marketing via the better and timely information supplied.

5/ Criteria for Proposals

Here is a simple but effective criteria by which you can evaluate your proposal before submitting it to management.

- Is the proposal written concisely and in language free of computer/technical jargon so that management can easily understand it?
- Does the proposal describe what the new/modified system will do?
- Do the conclusions flow logically from the observations?
- Is the present system (if there is one) referenced and related to the proposal?
- Are the charts, figures and tables uncomplicated and easy to follow?

- Is the method of approach, i.e., the method by which you intend to organize and approach the project, and go from one phase to another, well defined and logically tied together?
- Is the tone of your proposal positive without over-simplifying the difficult or potential problem areas?

SAMPLE APPLICATION

Proposal for

AUTOMATIC CREDIT CARD VERIFICATION SYSTEM

March 1974

Prepared by

Systems Group, Systems Department

BOULDER OIL COMPANY, INC.

CONTENTS

Section

Section I

INTRODUCTION

In the past year the use of stolen/lost and invalid credit cards has doubled over the previous year, costing the Boulder Oil Company a significant amount of money. The Control Department, therefore, requested the Systems Group to investigate the possibility of an automatic credit card verification system to be installed at each of Boulder's 250 service stations in Northern California.

Consequently, in accordance with Service Request No. 495 and the subsequent approved and signed assignment, the Systems Group for the past six months has been investigating the feasibility of a system that would meet the following requirements:

- Provide automatic credit card verification by putting individual service stations "on-line" with the Company's central computer in San Francisco.
- Identify stolen/lost or invalid credit cards.
- Be compatible with currently issued credit cards.
- Be simple in operation in order to require minimal training to operate the system.
- Offer significant operating advantages to the service station attendants. This is a very important point, since without the support of the operating people the best system would fail.

Analysis of the problem proceeded with the understanding that a proper solution depended upon a balance among such factors as cost, speed, modularity and schedule.

The objectives of our preliminary studies were:

- Examine Boulder Service Stations' existing method for the Company's credit cards processing.

- Design and develop a system that permits automatic credit verification at the point of sale.
- Survey commercial communication facilities and evaluate software packages and/or hardware that would best serve the contemplated system.
- Estimate the possible cost of implementing such system, including cost of new equipment, conversion and training of employees.
- Compare the cost of operating the new system with the old, and estimate the reduction in fraudulent credit card loss over a period of five years.
- Justify the new system by using return-on-investment (ROI) technique, the systems group being quite aware that management very seldom approves a proposed system unless it is fully justified.

The results of the study are very encouraging. The Systems Group has determined that a system as required is not only possible but feasible. The system we recommend is similar in approach to the credit verification system being used successfully by large department stores at the point of sale. Moreover, the Systems Group has checked, and the price of the recommended TRB miniterminals are most reasonable. Here too, the service station attendant/sales personnel keys in the customer's account number on a simple, adding machine-like TRB miniterminal, and within seconds the affirmative or negative response comes through. The proposed minicomputer (PDP-11) contains a list of account numbers for which credit will not be extended. This list is referred to as a "Negative list." If the response is affirmative, the price of the merchandise and/or service is keyed in and is then posted to the customer's account.

Communications is possible through five keys that light up on the miniterminal when the customer's account number is keyed in. Actually, there are six alphabetical keys on the miniterminal in addition to ten numerical keys (from 0 to 9), but the sixth key is the Attention key. The latter is depressed to contact the host computer's Central Processing Unit (CPU). The five alphabetical keys are: OK, RE-ENTER (in case the attendant did not enter the account number correctly), STOLEN, INVALID (credit overextended), and REFER.

Aside from eliminating improper credit cards and automatically updating the customer's account, this system will enable management to have truly up-to-date figures of the account receivable, since every night via batch-processing all the charges sent in and stored in the on-line storage of the proposed minicomputer (PDP-11) are used to update the Account Receivable Master File and the Service Stations Master File. In the morning a Management Report, in addition to a Daily Journal and an Exception Report, are printed of all of the previous day's transactions of every one of Boulder's 250 service stations.

The Systems Group's preliminary study has determined that the Company's existing host computer, IBM System 360, Model 50, currently used for account processing and personnel payroll, could be adapted for automatic credit verification and inputting of accounts receivable on a time-sharing basis. That is, the front end via the PDP-11 minicomputer could be used for the on-line inquiry retrieval system, while the back of the host computer could be used for batch processing.

As an added feature, the Systems Group recommends to utilize visual aid by having each one of the 250 Boulder Service Stations post a large sign saying: THIS SERVICE STATION USES AUTOMATIC CREDIT CARD VERIFICATION SYSTEM. Such signs would effectively deter certain individuals who make a practice of using stolen, lost or invalid oil company credit cards from trying to buy anything from any of the Boulder Service Stations. And it would save the attendant embarrassment and possible trouble, because he would have to ask the person for cash if his credit has been overextended, or if the credit card is a stolen one.

Supporting information may be found in the Discussion Section of the proposal, as well as work flowchart of the present and proposed system, description of new equipment needed, schedule of development (activity chart), estimated benefits and growth potential.

Section II

DISCUSSION

Under the present system (Figure A-1), the receipts for service or merchandise are mailed in daily to the Boulder Data Processing Center in San Francisco. At the Center, the receipts are processed through several clerks, and accounts are checked manually against Lost/Stolen and Invalid lists before the charges with valid account numbers are key-processed and then input to the Company's host computer to post these amounts in individual customer's account, and subsequently update the Account Receivable Master File and the Service Stations Master File. While copies of these lists are sent out regularly to all Boulder Service Stations, management recognizes that the attendants cannot very well take time out to check every customer's credit card against the lengthy "Negative" list, especially during the two busy times of the day, morning and evening, when most of the improper credit cards are being used.

Under the new system (Figure A-2), instead of asking for the credit card *after* selling gas or automobile parts, or performing repair service, the attendant asks for it *prior* to selling merchandise or providing service. The attendant keys in the credit card account number for verification (no need to change the currently issued credit cards; the proposed system is compatible with the Company's credit card account numbers), and within a few seconds the attendant receives one of five responses via the on-line inquiry system that checks the account number against an abstract file that the host computer updates nightly by batch processing. The five responses on the terminal are: OK, STOLEN, INVALID, REFER, RE-ENTER. The last response lights up when an account number is entered incor-

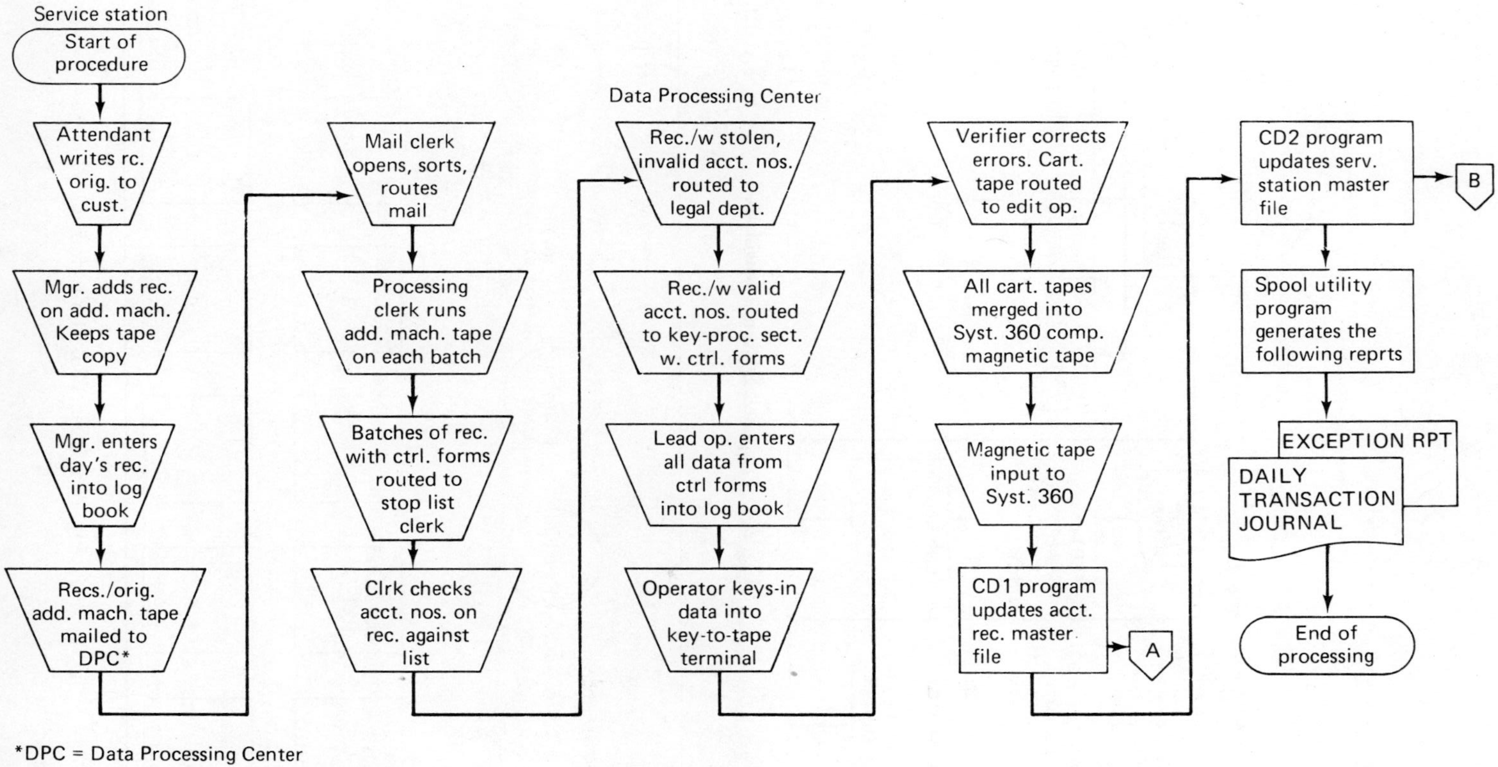

Fig. A-1 Sample Application—Present System Flowchart

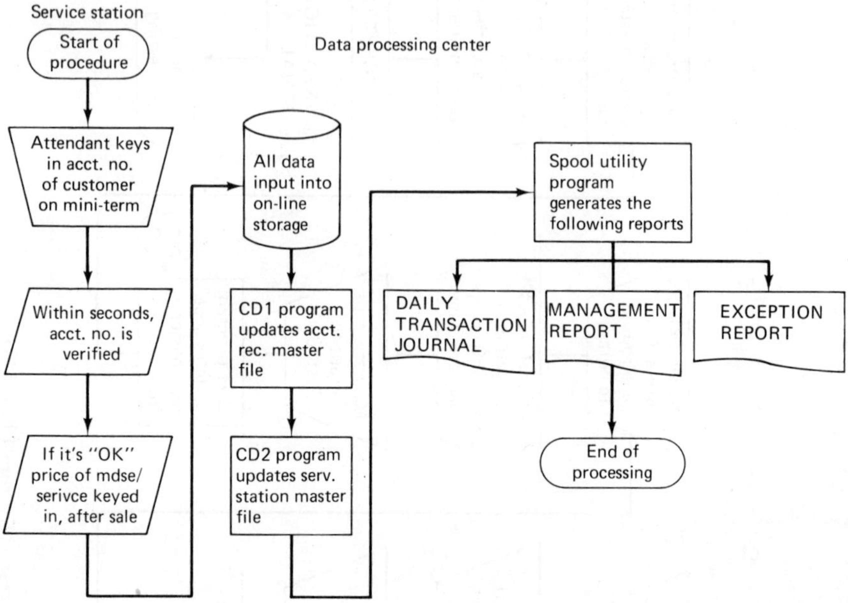

Fig. A-2 Proposed System Flowchart

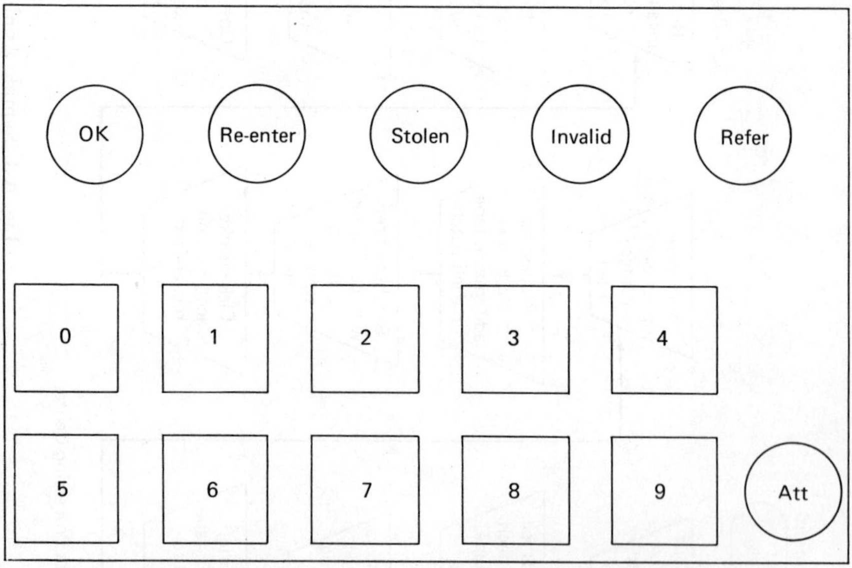

Fig. A-3 Basic TRB Miniterminal

rectly, such as too many or too few digits. In addition, there is an ATTENTION key (via which the attendant connects the terminal to the access controller, the PDP-11 minicomputer and on to the host computer) and ten numerical keys, 0 to 9, for keying in the account numbers.

Based on ROI (return-on-investment) analysis (definitive figures are in the Cost Analysis Section), the Systems Group estimates that the cost of TRB miniterminals (Figure A-3) to be installed in each of the 250 Boulder Service Stations, and the PDP-11 minicomputer to be installed at the Company's data center to control access to the Host computer, would be recovered within five years or less through the elimination of the use of lost/stolen and invalid Boulder Oil Credit Cards.

The preliminary study has established that the price of the TRB miniterminals are quite reasonable, especially since commercial telephone lines would connect the service stations with the central computer. Thus, no special communications links would be required; not for this phase, at least. In addition, this terminal system could be used in the future for entering orders for merchandise, and updating and controlling inventory at the service stations with very little modification to the original system.

Hardware configurations and software features of the proposed system are detailed in the Technical Approach Section.

In summary, the Systems Group is confident that the project can be performed as outlined in the Schedule of Development/Activity Chart (Figure A-4), and as described below:

A/ GENERAL DESIGN

This phase of the project consists of further macro level design of the system, founded on solid establishment of the system requirements. The design is based upon the preliminary specifications as outlined in this proposal.

B/ DETAIL DESIGN

This phase includes the development of all segments within the system, control sequence, input/output interfaces and so on, and the detailed logic diagramming of these routines.

C/ PROGRAM SPECIFICATIONS

In this phase all the program specification requirements are established, including the system output/input structures, the system processing structure and the system data and storage structures.

D/ PROGRAMMING

In this phase all programming, debugging and systems testing is performed.

E/ INSTALLATION/CONVERSION

In this phase the system is installed, the old system is converted to the new system and the service station attendants are trained to use the new system efficiently.

F/ DOCUMENTATION

A work book will be maintained during the full term of the project. Documentation of the Automated Credit Card Verification System will consist of system material which will contain all the information necessary for maintenance and/or modification to the system, and training/operations material, which will include a User Manual, a Training Manual, and a Service Manual. These last two manuals will be designed and written to be as legible and self-explanatory as possible, to minimize training time and eliminate attendant/operator confusion.

G/ DELIVERY

At the completion of this phase the proposed system will be in operation.

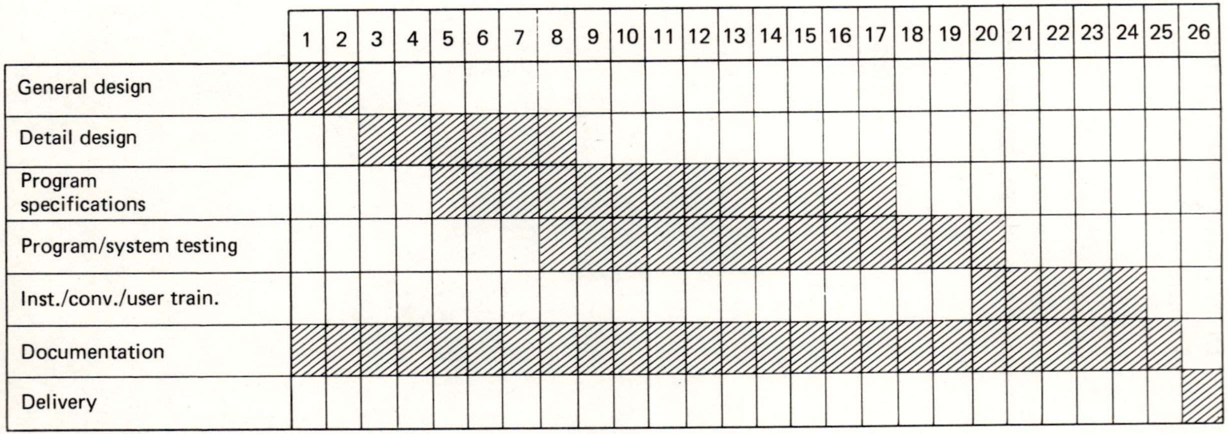

Fig. A-4 Activity Chart

Chapter 2

Development Phase

This chapter describes the approach to be used in performing each task within the development phase to define the existing and proposed system. Since this phase is carried out only after management has approved the proposal, it marks the actual start of the full-scale system development. This being the case, a thorough investigation of the present system (if there is one) and a definition of the proposed system is the logical activity to effect a functional, structured system.

It is also at this stage that you and the user/management really begin to work together. Accordingly, keep in mind one of the most important rules in systems analysis: Don't work *for* them; work *with* them.

This phase as well as the next two phases have to be planned carefully so that they are accomplished within the time stated in your proposal's activity chart (Figure A-4). However, if you have done a thorough problem analysis by getting the "big picture;" analyzing the service request, tracing the flow of work in the particular system via a flowchart and conducting in-depth interviews, you already have most of the information you need in this section for your project. Now it is only a matter of pulling all information together and presenting it in an attractive and legible form.

The development phase comprises three tasks: A/ system requirements, B/ general system design and C/ checkpoint/sign-off.

A/ SYSTEM REQUIREMENTS

The purpose of the system requirements task is to define the existing system via a system study. This is necessary so that you understand how and why the present system operates. Analysis of the present system also serves to ensure that the new or revised system (to be designed) will avoid the inadequacies of the present system and truly meet the requirements of the user/management.

To get a thorough knowledge of the present system, review the interviews you have conducted and documented in the previous phase. Also, review how the work or the system data flow from beginning to end through one or many departments. In other words, analyze the procedures and the forms (used in the procedures) in relation to the flow of work.

If the present system is computerized, and if it has any documentation, a study of those documents is necessary. If the documentation is out of date it should be upgraded, *if* the project time and budget allow it.

The system requirements task consists of eight well-defined steps: 1/ analysis of procedures, 2/ analysis of forms, 3/ analysis of job functions, 4/ analysis of inputs, 5/ analysis of outputs/ reports, 6/ analysis of files, 7/ analysis of controls and 8/ summary of present system.

1/ Analysis of Procedures

If there are any written procedures of the present system get a copy of them. After summarizing the particular procedure on the Procedure Analysis form (Figure 2-1), compare the activity with your work flowchart. Quite often your work flowchart will disclose significant deviations between what the written procedures and other documentation indicate the system is supposed to do, and what it is actually doing.

If there are no written procedures of the present system, or if they are inaccurate, you should document the procedures as it is stated on your work flowchart.

CROCKER NATIONAL BANK
DATA PROCESSING AND SYSTEMS ADMINISTRATION

PROCEDURE **ANALYSIS** Page

[X] INPUT [ ] OUTPUT

PROJECT NO.
TDS1234

PROJECT	PREPARED BY	DATE PREPARED
TIME DEPOSIT SYSTEM	AN ANALYST	3-27-74
DEPARTMENT	**DEPARTMENT SUPERVISOR**	**APPROVED BY**
BANK OPERATIONS	A SUPERVISOR	

DOCUMENT NO.	COPY ATTACHED	FORM TITLE	FORM NO
02-500	[X] YES [ ] NO	DEPOSIT TO SAVINGS ACCOUNT	02-500

FORM PURPOSE

To handle credit (deposit) entries to Bank savings accounts.

	DUE IN	DUE OUT
SCHEDULE	8:00 a.m. Daily (from DPC)	5:00 p.m. Daily (to the DPC)
SOURCE DEPARTMENT(S)	All branches with automated savings.	

PROCESS AND EDIT STEPS	PERFORMED BY
1. MICR-encode account number and amount	MICR-ENCODER
2. On an adding machine list and total all deposits	TELLER
3. Enter branch name, date, and transaction code (7) on the listing.	OPERATIONS SUPERVISOR
4. Place form 06-500, MIDAS Header, in front of each batch	TELLER
5. MICR-encode batch total on form 06-049, MIDAS Trailer	MICR-ENCODER

CONTROL	VOLUME				
	PERIOD	FREQ.	MIN.	AVG.	MAX
Tape of all deposit amounts and total	Sep-June	Daily	500	700	1000
Batch total on MIDAS Trailer	Jul-Aug	Daily	400	600	900

DISTRIBUTION

DEPARTMENT NAME	NO./COPY	FILE REFERENCES
Check Processing and Accounting - DPC	Original	

COMMENTS

The DPC is responsible for processing the deposit tickets(either keypunch or MICR capture), and returning the document to the branch with other transactions, exception listings, etc., of the day.

Fig. 2-1 Procedure Analysis Form (Courtesy of Crocker Bank)

The procedures (this applies to both the present system and the new system to be designed) should describe:

- Which employee is doing what work and, if time is important in the procedure, how long it takes to perform that job. In other words, the starting and ending time of the procedure has to be stated.
- The title and number of forms being used in the procedure.
- How the procedures are controlled, i.e., are there check/controls at critical points in the procedures?

For techniques on how to write functional, legible procedures (with examples) see Part 2 of this reference manual.

2/ Analysis of Forms

Since forms used in system documentation are very important, secure a copy of every form being used in procedures, input, output, files and controls of the present system.

Analyze each form in regard to:

- Design. Is it simple and functional or confusing and unnecessarily complex?
- Possible duplications. If there are duplications is it possible to combine those forms?
- Use of the form in the system.
- Number of copies. Are any forms made out in too many copies, wasting paper and increasing cost?
- Frequency of usage of the form.

3/ Analysis of Job Functions

Review the interviews you have conducted with the clerical and operator type personnel in the previous phase;

compare them to the specific procedure the employee is performing and document these functions on the Job Functions form (Figure 2-2). This form should be filled out as concisely as possible, and given to the particular employee. Ask him/her whether his functions as you have documented them are correct, and solicit comments relating to the work. At the end of this step you will be able to spot overlapping and duplicated functions, trouble spots, errors and procedural problems.

4/ Analysis of Inputs

This is the step where you may find the reason for unsatisfactory output: incorrect or inadequate input.

Input documents (Figure 2-3) should be analyzed in terms of the following:

- Is the input form to the present system specific and self-explanatory or is it vague and confusing?
- Is there too much information on the input document, i.e., is it trying to give too much data for too many purposes?
- Is the input document made out in too many or too few copies?
- How is the document routed to other functions?
- Is the input document distributed to too many people?
- Are there any controls to prevent the document's loss?
- Is there a method for finding a missing document?
- Is there a method for checking the accuracy of information on the input document?
- If the present system is manual, can the input form be used for the proposed computerized system?
- What is the cost of the present input form? Is there a less expensive method to produce it?

CROCKER NATIONAL BANK
DATA PROCESSING AND SYSTEMS ADMINISTRATION

JOB FUNCTIONS

[X] EXISTING ☐ NEW

Page

PROJECT	PREPARED BY	PROJECT NO
TIME DEPOSIT SYSTEM	AN ANALYST	TDS1234
DEPARTMENT	**DEPARTMENT SUPERVISOR**	**DATE PREPARED**
CHECK PROCESSING AND ACCOUNTING	A SUPERVISOR	4-3-74
NAME	**POSITION**	**APPROVED BY**
John D. Jackson	CONTROL CLERK	

FUNCTION NO.	% OF TIME	HOURS PER DAY	HOURS PER WEEK	HOURS PER MONTH	VOLUME	FUNCTION DESCRIPTION
01	25	2			1000 Trans-actions	Receive Savings transactions from the branches; sort to account number within branch, within transaction type and batch.
02	25	2			1000 Trans.	Post balances to control sheet by branch.
03	12	1				Post control sheet to totals produced by the branches.
04	-	-				Forward transactions to computer room.
05	20		8			Balance totals of Balance & Transaction Register to control sheets.
06	17		7			Correct all unposted items.
TOTAL	100	5	15			

COMMENTS

1) This job is performed under tight scheduling.

2) There is some duplication of effort (functions 1 and 2) in that the branches also perform these same functions.

Fig. 2-2 Job Functions Form (Courtesy of Crocker Bank)

CROCKER NATIONAL BANK
DATA PROCESSING AND SYSTEMS ADMINISTRATION

INPUT DOCUMENT
DESCRIPTION

Page

PROJECT			PREPARED BY	PROJECT NO / DATE PREPARED
TIME DEPOSIT SYSTEM			AN ANALYST	TDS1234 / 3-14-74
DEPARTMENT			DEPARTMENT SUPERVISOR	APPROVED BY
BANK OPERATIONS			A SUPERVISOR	
DOCUMENT NO	COPY ATTACHED	FORM TITLE		FORM NO
05-800	☒ YES ☐ NO	NAME AND ADDRESS NOTICE		05-800

FORM PURPOSE

To enter name and address information for savings accounts into the Uni Name & Address System.

PRESENT INPUT FORM

Preprinted form

PROPOSED INPUT FORM

OCR form

PROCESS AND EDIT STEPS	PERFORMED BY
Complete form using OCR font on selectric typewriter	Typist
Retain one copy in branch files; forward one copy to DPC	Op. Supervisor
Batch by office number on each Friday; submit for OCR capture	Control Clerk DPC
Perform OCR capture; check listing for rejects. Keypunch all rejects; submit with OCR tape to N/A system for computer processing.	Control Clk. DPC
Return N/A listing& card file updates to branches.	Control Clk DPC

CONTROL	VOLUME				
	PERIOD	FREQ.	MIN.	AVG.	MAX.
OCR Capture Printout	Jan-Feb	Weekly	50	150	200

COMMENTS

Fig. 2-3 Input Document Description (Courtesy of Crocker Bank)

5/ Analysis of Output/Reports

Whether you are analyzing the output/reports (Figure 2-4) or the input, coordinate this activity with the information you have gained from your interviews and the data on your work flowchart. In addition, just as with procedures and forms, make sure that you have copies of all the present system's reports. Analyze each report with the following criteria in mind:

- Design. Is the report well designed for ease of understanding?
- Who is using the report?
- Are the users of the report satisfied with its form and the amount of information it gives?
- How is it prepared?
- How much time is required to prepare it?

At the end of your analysis of the present system's outputs/reports these are the problems you might find:

- Users of certain reports are dissatisfied because it is difficult to identify and use the reports. To the nonprogramming staff the reports are illegible and unnecessarily complex.
- Written input procedures for report users are unnecessarily technical and long to prepare.
- There are no standard written instructions for report handling and distribution.
- Most reports do not have descriptive titles and identification codes, thus usage by nontechnical staff is limited.
- There is no listing of all the reports the system outputs.
- There is no index of all the systems and applications the company is processing. In short, management does not know the status of all systems and applications.

CROCKER NATIONAL BANK
DATA PROCESSING AND SYSTEMS ADMINISTRATION

OUTPUT/REPORT DESCRIPTION

Page

PROJECT NO.
RCC064

PROJECT	PREPARED BY	DATE PREPARED
RETAIL CREDIT CARD PROCESSING	AN ANALYST	2-9-74
DEPARTMENT	**DEPARTMENT SUPERVISOR**	**APPROVED BY**
RETAIL CREDIT CARD	A SUPERVISOR	

DOCUMENT NO.	COPY ATTACHED	FORM TITLE	FORM NO.
D08	☐ YES ☒ NO	NON-MONETARY INPUT JOURNAL	Stock

FORM PURPOSE
Reference

SATISFIES REQUIREMENT OF: ☐ LAW/REGULATION ☐ BANK POLICY ☒ OTHER:

FREQUENCY		SEQUENCE: DATA ELEMENT	PAGE BREAK	SUM ON	MAJOR ↓ MINOR
DAILY					
WEEKLY	X	Account Number			
MONTHLY	X				
QUARTERLY					
DEMAND INQUIRY					
OTHER					

PAPER		PER PAGE PRINT LINE	VOLUME MIN.	AVG.	MAX.	OTHER OUTPUT MEDIA
PAPER SIZE	14 x 11					CARD
		HEADER LINES	1	1	1	
TOTAL LINES AVAILABLE/PAGE	66	DETAIL LINES	40	40	40	MAG. TAPE
		SUMMARY LINES	0	0	0	
		LINE TOTAL	0	0	0	PAPER TAPE
NO. OF COPIES	6	SPACE	20	20	20	
		SKIP	5	5	5	MICROFILM
PREPRINTED	No	TOTAL LINES	66	66	66	
		NO. OF PAGES	60	240	1440	OTHER

DISTRIBUTION

DEPARTMENT NAME	NO/COPY	FILE REFERENCE
Retail Credit Card	1 - 6	

COMMENTS

Journals are produced each week to include all valid daily transactions for the calendar week. These transactions are carried forward onto a cumulative file, and listed monthly for the four or five previous calendar weeks. The transactions for 26 calendar weeks are combined and a journal is produced on microfilm semi-annually.

Fig. 2-4 Output/Report Description (Courtesy of Crocker Bank)

6/ Analysis of Files

Thanks to some computer "authorities" a mystique has been created about "files." Yet a file is simply a collection of information. If the present system is manual, the files referred to can be a filing cabinet, a Kardex file or even a manilla folder in which all the records of the particular system are kept. Because it is essential that you know what information each file contains and why it is being maintained, whether it is a manual or computerized system, this step in system requirements is quite important.

Analysis of files in the present system can be performed by reviewing the file description forms (Figure 2-5). (There should be a File Description form made out for each file.)

An analysis of the files should include the following:

- Data in the files, records within those files and error tolerances permitted.
- Are files duplicated within the organization, i.e., are identical data maintained at several locations?
- Are the files organized so that they can be used efficiently for report preparation?
- What are the decision rules concerning inactive data within the records?
- What is the storage space and cost?
- Are there any existing procedures for file security and file protection?
- Is there a backup for each file?

Before finishing with this step you should know how frequently each file is referred to and how often it is updated. Analysis of the files also will serve to verify interviews and written procedures and indicate if there are any missing or incomplete data in your present system analysis.

CROCKER NATIONAL BANK
DATA PROCESSING AND SYSTEMS ADMINISTRATION

FILE DESCRIPTION

☒ EXISTING ☐ NEW

PROJECT	PREPARED BY	DATE PREPARED
TIME DEPOSIT SYSTEM	AN ANALYST	4-15-74
DEPARTMENT	**DEPARTMENT SUPERVISOR**	**APPROVED BY**
BANK OPERATIONS	A SUPERVISOR	

Page — PROJECT NO: TDS1234

FILE NO	FILE NAME
F-10	Savings Master Name and Address File

FILE PURPOSE
To provide addresses for statements to be mailed.

DOCUMENT REFERENCE
N/A on Master Tape

UPDATE INTERVAL: Daily	SECURITY: Tape Library
RECORDS IN FILE: 587,364	BACKUP: Signature Card In Branch
TRANSACTIONS: 3,500	RETENTION: One Year
SEQUENCE: Account Number	EQUIPMENT: Tape

ACCESS REQUIREMENTS

PURPOSE	To maintain correct name and addresses for statements each quarter.
HOW FAST	N/A
HOW OFTEN	Once each business day and quarterly.
BY WHOM	Update program
WHEN	Daily and quarterly
WHERE	SF DPC

CONTROL AND UPDATE PROCEDURE

Tape under standard control. Update made by OCR form originating from branches each time there is an address change.

COMMENTS

Fig. 2-5 File Description (Courtesy of Crocker Bank)

7/ Analysis of Controls

An analysis of controls should include a review of all of the previous steps. Specifically:

- What are the controls in the *procedures*? How effective are these controls?
- What are the controls for the input and output documents? How effective are these controls?
- What controls are exercised on the files insofar as the data they contain, frequency of usage and frequency of updating? How effective are these controls?

8/ Summary of Present System

When you arrive at this step it is a good idea to confirm and summarize all the facts you have gathered by describing in a nontechnical narrative how the existing system operates (Figure 2-6). The description should briefly explain the purpose and scope of the system, the source and nature of the input, the type of processing performed, the purpose and destination of the output and the controls exercised. While this description will serve you as a reference document throughout your project, it might also be useful in the last task of this phase—the checkpoint/sign-off task. Management may need to be reminded what the present system is doing and where the inadequacies are.

B/ GENERAL SYSTEM DESIGN

The purpose of the general system design task is to determine, organize and document the information needs of the user/management within the constraints of a responsive new system. Consequently, by compiling, constructing and preparing the general system flowchart, the general system narrative and an overview of the output/input, the systems analyst documents

CROCKER NATIONAL BANK
DATA PROCESSING AND SYSTEMS ADMINISTRATION

NARRATIVE SYSTEM DESCRIPTION
TITLE

PROJECT	PREPARED BY	Page / PROJECT NO. / DATE PREPARED
		RCC064
RETAIL CREDIT CARD PROCESSING	AN ANALYST	2-9-74
DEPARTMENT	DEPARTMENT SUPERVISOR	APPROVED BY
RETAIL CREDIT CARD	A SUPERVISOR	

Non-monetary input consists of three basic transactions for maintenance of the Cardholder Master File. 1. Addition of an account number, 2. Changes of data content for an account number, and 3. Deletion of an account number.

Daily:

These transactions are recorded by Retail Credit Card Department on OCR format documents and forwarded to SF DPC where this input is transcribed to magnetic tape. The tape is processed to reformat the records, and edit for erroneous or invalid data. Transactions containing errors are excluded from the two output files, and are itemized on an error listing for review and correction by Retail Credit Card.

Weekly:

The daily transactions which require Retail Credit Card issue materials are processed to produce a series of files representing the most current input for each account in card issue document format (mailers, embossing cards, etc.). These transactions and those others requiring journal entries are used to produce a journal listing of all valid transactions processed for the current week.

Monthly:

A journal listing is prepared representing the transactions itemized in each of the weekly journals for the current month, and the re-issue accounts from the monthly re-issue cycle.

Semi-Annually:

A journal listing is prepared representing the data itemized in each of the monthly journals for the prior five months plus the current month. This journal is then recorded on microfilm.

Fig. 2-6 Narrative System Description (Courtesy of Crocker Bank)

up to this point the manual and/or automated operations necessary to satisfy the established information requirements.

To ensure the quality of the necessary information produced, define and perform the proposed activities in a logical order. The result should mirror a careful consideration of the user's needs, basic system techniques, data processing methodology, optimum use of available resources and an appropriate degree of system flexibility for possible future expansion.

In the preceding activities, the user's needs and design constraints were identified and documented. In this task and the detailed system design task, as the architecture of the new system is finalized, it is necessary to impose some or all of those constraints and any additional ones that become apparent.

A sign-off by the user/management at checkpoint—the last step in this phase—is, in effect, a contract between you and that person. That is, it is a document of the user/management's approval of the proposed system; it is your promisory note to deliver a product that meets the agreed-upon specifications within the stated time.

The general system design task comprises three precise steps: 1/ general system flowchart, 2/ general system narrative and 3/ general output/input.

1/ General System Flowchart

There is much more to the general system flowchart than just graphically describing from an overview level the sequence of processing of the data within the system. Although this flowchart's primary function is to illustrate how the source inputs of the system are manipulated and combined with the data files to produce the required outputs, it is really the blueprint for the architecture of the new system. Consequently, you will have to review all of your data from the previous phases and add any new data you may have picked up to ensure that you have defined exactly the requirements of the new system. Once you have identified and defined the user/management needs, you can utilize your creativity and knowledge by designing a system that responds to the stated requirements within the defined constraints.

To be able to draw a fairly accurate General System Flowchart, prepare the following based on the data collected in your documentation workbook. Most of these documents will also serve for future reference.

a/ *Output/Reports*

- Define the contents of your output/report so that they are consistent with the proposed use of the documents.
- Design the report format so that the headings, descriptions and other information are self-explanatory.
- State the frequency of the reports (daily, weekly, monthly) and define how many copies of each report are to be printed after you have checked the number of individuals who have an actual need for the particular output/report.
- Summarize the above in the output/report form (Figure 2-7).

b/ *Control Requirements*

- Define how in your proposed system the inactive data will be removed from the files.
- State how the erroneous data will be corrected.
- Establish a method for handling classified material (if there is any).
- Summarize the above control requirements in the control summary form (Figure 2-8).

CROCKER NATIONAL BANK
DATA PROCESSING AND SYSTEMS ADMINISTRATION

OUTPUT/REPORT DESCRIPTION

Page
PROJECT NO: TDS1234

PROJECT	PREPARED BY	DATE PREPARED
TIME DEPOSIT SYSTEM	AN ANALYST	4/2/74
DEPARTMENT	DEPARTMENT SUPERVISOR	APPROVED BY
BANK OPERATIONS	A SUPERVISOR	

DOCUMENT NO	COPY ATTACHED	FORM TITLE	FORM NO
D01	☒ YES ☐ NO	DAILY BALANCE AND TRANSACTION REGISTER	07-500

FORM PURPOSE

To provide branch offices and subscribers with a register of account balances, status, and transaction activity on a daily basis.

SATISFIES REQUIREMENT OF: ☐ LAW/REGULATION ☒ BANK POLICY ☐ OTHER:

FREQUENCY		SEQUENCE			
		DATA ELEMENT	PAGE BREAK	SUM ON	
DAILY	active accts				
WEEKLY	complete	SUBSCRIBER	X	X	MAJOR
MONTHLY		BRANCH	X	X	↓
QUARTERLY		ACCOUNT			
DEMAND INQUIRY					
OTHER					MINOR

PAPER		PER PAGE PRINT LINE	VOLUME MIN.	VOLUME AVG.	VOLUME MAX.	OTHER OUTPUT MEDIA
PAPER SIZE	11 × 14	HEADER LINES			1	CARD
TOTAL LINES AVAILABLE/PAGE	88	DETAIL LINES			37	MAG. TAPE
		SUMMARY LINES			-	
		LINE TOTAL			1	PAPER TAPE
NO. OF COPIES	1	SPACE			49	
		SKIP				MICROFILM
PREPRINTED	Yes	TOTAL LINES			88	
		NO. OF PAGES		800	1200	OTHER

DISTRIBUTION

DEPARTMENT NAME	NO/COPY	FILE REFERENCE
BRANCH OFFICES	ORIGINAL	

COMMENTS

Summary total boxes at the bottom of each page has been added for Inactive Balance and Inactive Count.

Fig. 2-7 Output/Report Description (Courtesy of Crocker Bank)

CROCKER NATIONAL BANK DATA PROCESSING AND SYSTEMS ADMINISTRATION	TITLE SUMMARY OF CONTROLS		Page PROJECT NO. TDS1234
PROJECT TIME DEPOSIT SYSTEM		PREPARED BY AN ANALYST	DATE PREPARED 4/74
DEPARTMENT BANK OPERATIONS		DEPARTMENT SUPERVISOR A SUPERVISOR	APPROVED BY

I. CONTROLS TO BE EXERCISED IN BRANCH OFFICES

1. Enter and verify on key-to-tape terminals all monetary transactions. Separate all credit and debit items. Erroneous data is corrected and re-entered.
2. Check digits across the system via the terminal.
3. Batch totals on trailers.
4. Totals on daily BTR and Inactive Accounts balanced to original transmittal tapes.
5. Check Entry letters sent to the DPC and received from the DPC.

II. CONTROLS TO BE EXERCISED IN THE DATA PROCESSING CENTER

1. Verify Batch and BTR totals against the transmittal tapes.
2. Capture program error listings supplied by:
   a. MID capture program SARIA
   b. CID capture program SARIB
3. Use Unposted Items Listing provided by Posting program (SARO) to create for re-entry the corrected items for the following day.
4. Document and standardize control procedures.

Fig. 2-8 Summary of Controls Form (Courtesy of Crocker Bank)

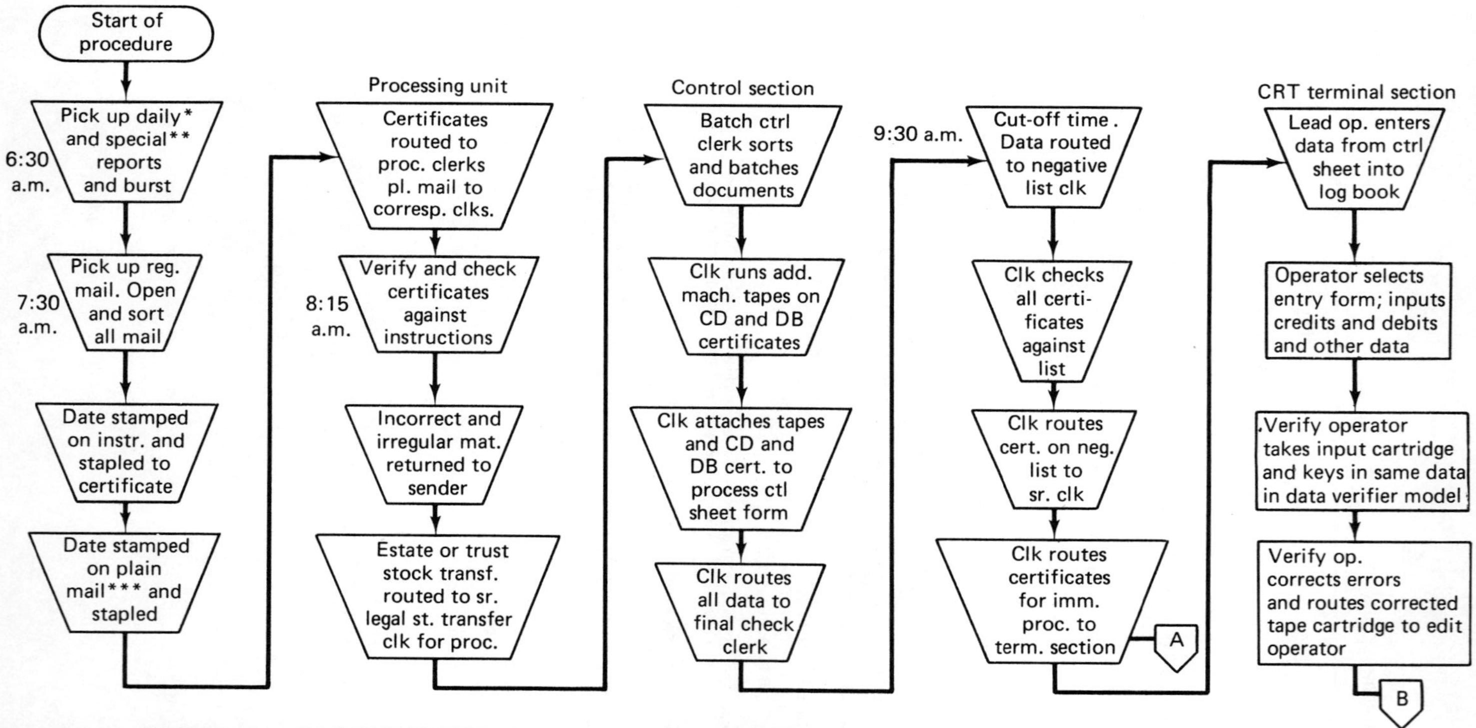

Legend: *Daily Reports: DAILY TRANSFER JOURNAL, TRANSACTION VALIDATION LISTING, RECORD OF CHANGES
**Special Reports: SHAREHOLDERS LISTING, DIVISION AND PROXY EXTRACTIONS, ACCOUNT ANALYSIS
***Plain Mail: Correspondence re.; change of address, name, or soc. sec. no., stop-payment notice, etc.

Fig. 2-9 Stock Transfer Work Flowchart

c/ *Work Flowchart*

- Draw a work flowchart of how the work in your proposed system cycle flows through a department or series of departments.
- Illustrate via your flowchart the complete processing flow of one system cycle by the frequencies of the necessary functions. In other words, there should be a separate flowchart for the daily processing, weekly processing and monthly processing of the system cycle.
- Be sure you include in your work flowchart the decision rules to be employed, exception functions and controls (Figure 2-9).

d/ *Data Structure*

- Define the problems and state your recommended solutions (Figure 2-10); if possible, visually illustrate how you intend to effectively structure data in the new system (Figure 2-11).

e/ *Input*

- Identify and define as much as possible the sources, types and volumes of input documents needed for the new system. Detailed input data description is covered in the solution phase.
- Summarize your requirements of the input in the general input description form (Figure 2-12).

f/ *Design Constraints*

- Define the schedule/timing constraints of the new system if the basic design is dependent upon any schedule changes to satisfy the user's requirements.

PERSONNEL SYSTEM DATA STRUCTURE

Problem Definition

(1) Records tend to be excessively long
(2) Redundant data are maintained in multi-locations
(3) Unrestricted access to confidential data

Recommended Solution

(1) Limit length of records
(2) Eliminate redundant data in multi-locations
(3) Restrict access to specific elements of data
   - Payroll
   - Medical
   - Personal
(4) Have alternate methods of access—randomly
   - By employee number or name
   - By skills (multiple)

Fig. 2-10 Personnel System Data Structure

- Document any policy and/or law/regulation constraints.
- Summarize the above in the design constraints form (Figure 2-13).

When you have completed the above documents, you are ready to prepare the general system flowchart (Figure 2-14) using standard flowchart symbols (Figure 2-15). You may want to supplement your general system flowchart with a processing schedule flowchart such as Figure 2-16. Organize your flowchart so that all processing done on the same time scale is shown on the same page; there should be

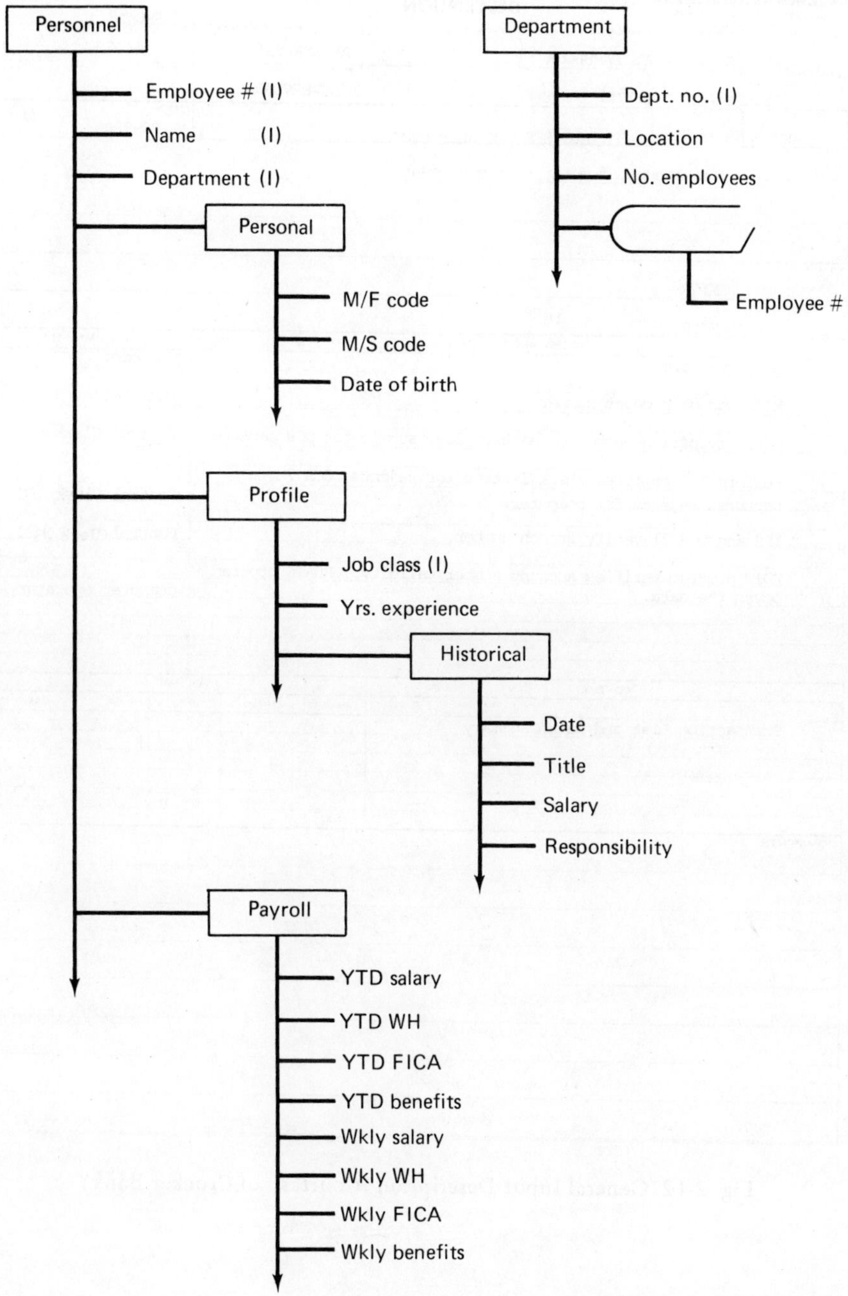

Fig. 2-11 Personnel System New Data Structure

CROCKER NATIONAL BANK
DATA PROCESSING AND SYSTEMS ADMINISTRATION

GENERAL INPUT DESCRIPTION

Page

PROJECT: SAVINGS AND BONUS PASSBOOK	PREPARED BY: AN ANALYST	PROJECT NO: SBP078
		DATE PREPARED: 4/74
DEPARTMENT: SYSTEMS AND PLANNING	DEPARTMENT SUPERVISOR: A SUPERVISOR	APPROVED BY:
DOCUMENT NO: 05-800 — COPY ATTACHED: ☒ YES ☐ NO — FORM TITLE: SAVINGS DEPOSIT SLIP FORM		FORM NO: 05-800

FORM PURPOSE

To add an amount of money to the account.

PRESENT INPUT FORM

MICR

PROPOSED INPUT FORM

MICR

PROCESS AND EDIT STEPS	PERFORMED BY
Complete form	Customer
MICR encode account number	Bank Teller
Batch by office number on each Friday, submit for OCR capture	Control Clerk DPC
Perform IPS capture, check listing for rejects, OCR 71 re captures rejects for computer	Control Clerk DPC
IPS and OCR 71 verify Branch number.	Control Clerk DPC
EDIT program verifies account number validity, UPDATE program posts the data.	computer operator

CONTROL	VOLUME: PERIOD	FREQ.	MIN.	AVG.	MAX.
Transaction List and Batch Summary					

COMMENTS

Fig. 2-12 General Input Description (Courtesy of Crocker Bank)

CROCKER NATIONAL BANK
DATA PROCESSING AND SYSTEMS ADMINISTRATION

Page

TITLE: DESIGN CONSTRAINTS		PROJECT NO: RCC064
PROJECT: RETAIL CARD PROCESSING	PREPARED BY: AN ANALYST	DATE PREPARED: 4/74
DEPARTMENT: Retail Credit Card	DEPARTMENT SUPERVISOR: A SUPERVISOR	APPROVED BY:

I. Schedule/Timing Constraints

A. Daily

1. Non-monetary input to be delivered to SFDPC by 4:30 p.m.
2. Error listings to be delivered to RCCD by 8:30 a.m. next day
3. Transactions to be delivered to WSBA before 12:00 p.m.

B. Weekly

1. Journal listing to be delivered to RCCD by 8:30 a.m. Monday
2. Issue documents to be delivered to RCCD by 8:30 a.m. Friday

C. Monthly

1. Journal listing to be delivered to RCCD by 8:30 a.m. 1st Monday

D. Semi-Annually

1. Journal listing tape to be delivered to COM room by 1st week of Jul and Jan
2. Journal microfilm to be delivered to RCCD by 2nd week of Jul & Jan

II. Policy Constraints

No constraints in this area at this time.

III. Technical Feasibility

No constraints in this area at this time.

IV. Law/Regulatory

No constraints in this area at this time.

Fig. 2-13 Design Constraints (Courtesy of Crocker Ban

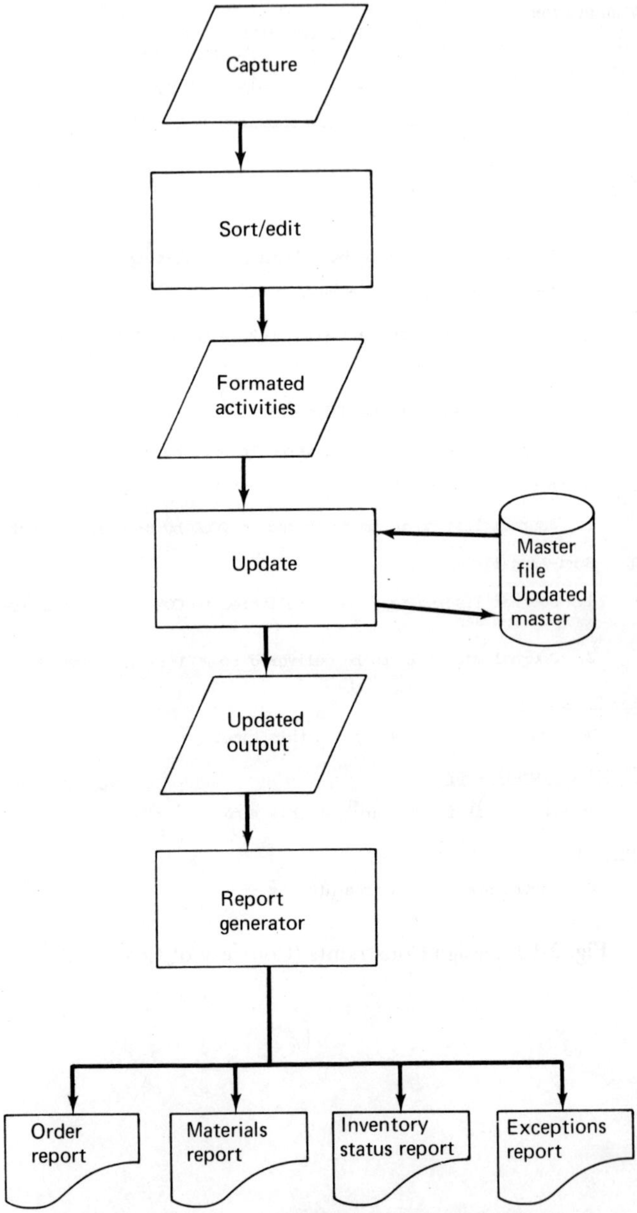

Fig. 2-14 General System Flowchart

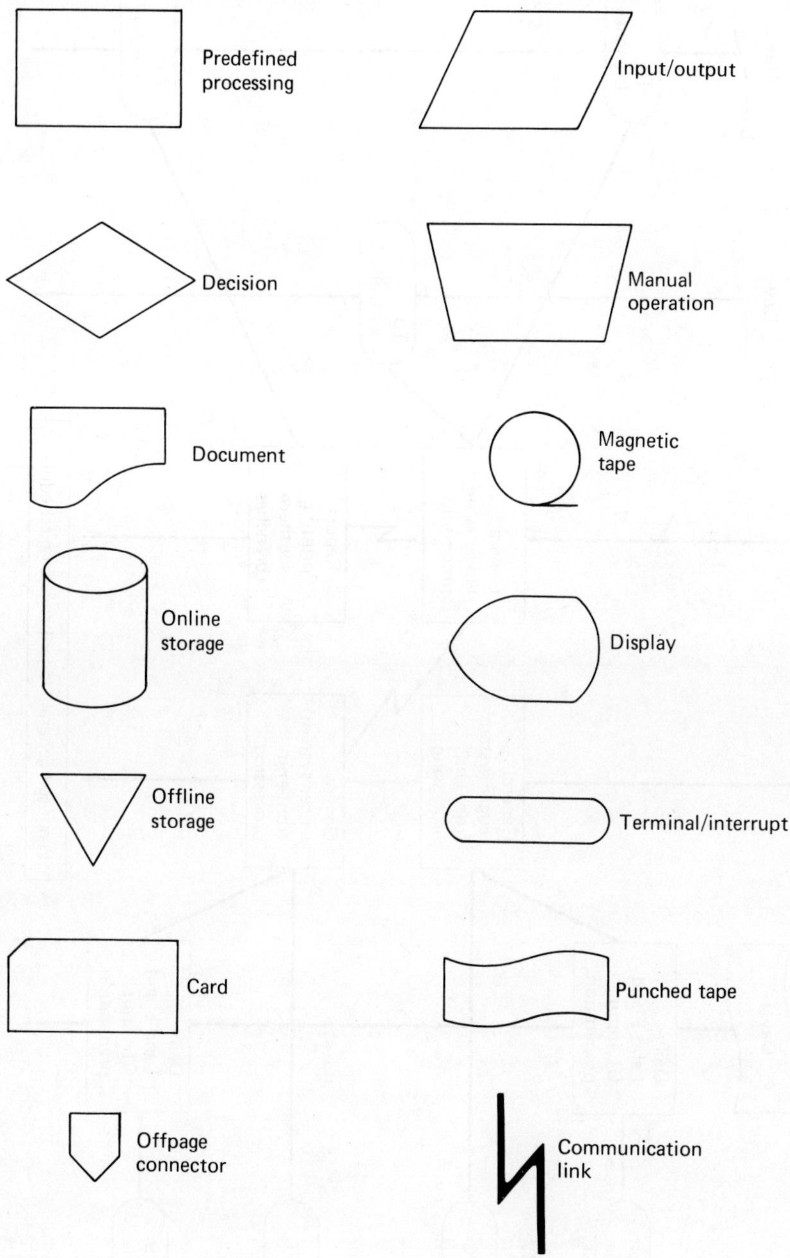

Fig. 2-15 General System Flowchart Symbols

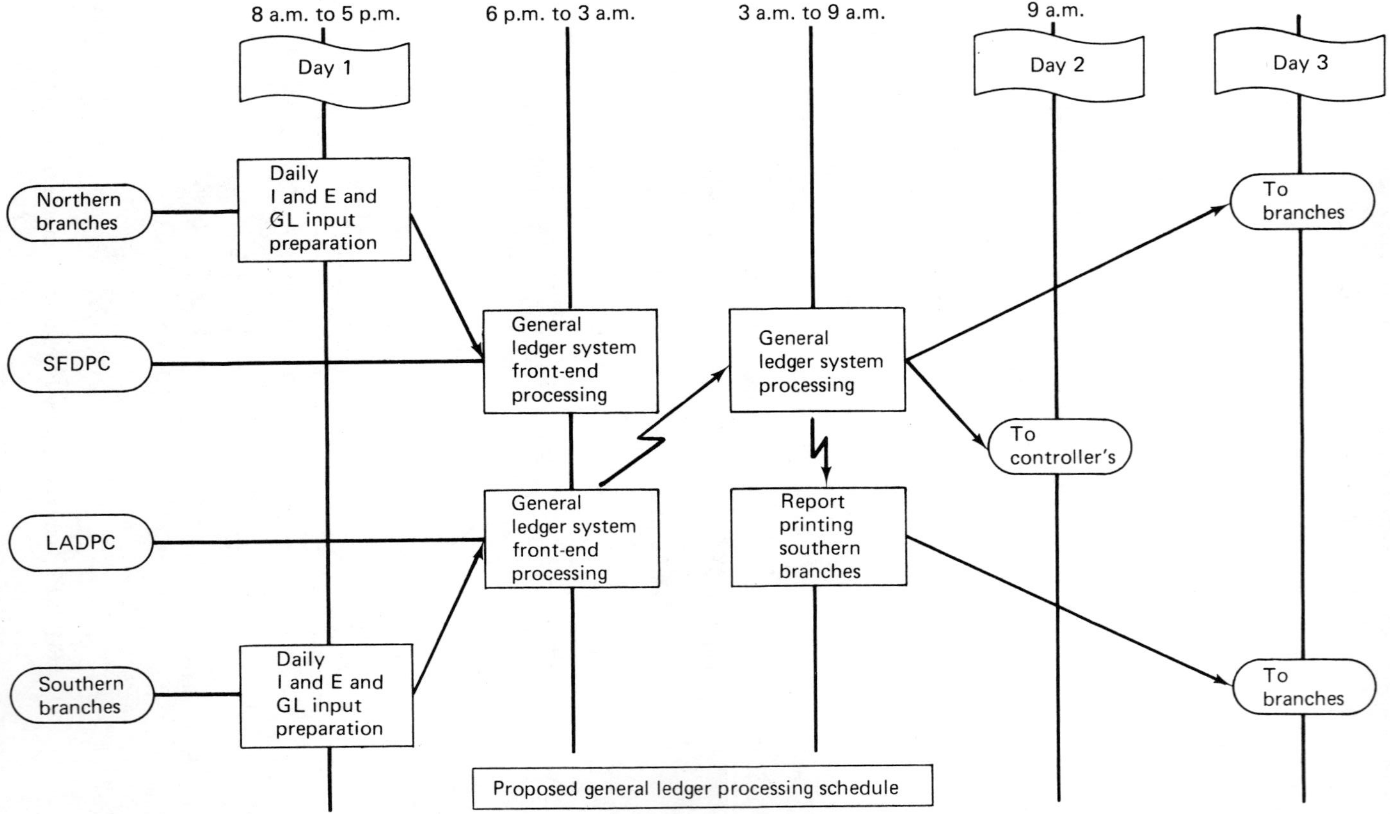

Fig. 2-16 Processing Schedule Flowchart

allowance for error processing and controls. Design your system so that the operation is as simple as possible.

After your general flowchart is finished, check to ensure that all of the required input and output files are included; that all captions and headings are indicated; that each input, process and output have been given the same title throughout the flowchart; that data audit trials have been included; that the frequency of the run is noted; and last but certainly not least, that the flow of system procedure is easy to follow and understand. In other words, keep your general system flowchart as simple as possible. Complicated design or flowchart do not make for efficient systems.

2/ General Systems Narrative

In the general system narrative (Figure 2-17) you describe briefly and legibly the intention, scope, features and benefits of the new system. Be sure that you include the input, the type of processing and the output. Give a list of the reports together with a short summary of their content. Since this document will be read (at the end of this phase) by user/management, some of whom may not be computer-type people, keep technical terminology and computer jargon to a minimum.

3/ General Output/Input

At this stage of your system development project you explain to the user what the output of the new system will be (this is what user/management is really interested in) and the input the user is required to provide to produce the required output. The nontechnical narrative of the output/input you write is based on the output/reports

Page

CROCKER NATIONAL BANK DATA PROCESSING AND SYSTEMS ADMINISTRATION	TITLE GENERAL SYSTEM NARRATIVE		PROJECT NO TDS1234
PROJECT TIME DEPOSIT SYSTEM		PREPARED BY AN ANALYST	DATE PREPARED 4/74
DEPARTMENT BANK OPERATIONS		DEPARTMENT SUPERVISOR A SUPERVISOR	APPROVED BY

A. GENERAL DESCRIPTION

The Time Deposit System will process all time deposit functions: Savings, Savings and Loan Savings, Certificates of Deposit, Christmas Club. Each of these subsystems is described in further detail on separate narrative forms. The system, by utilizing an 'option' concept on the subscriber and account grouping levels, is able to provide flexibility in such areas of processing as:

interest rate	interest calculation method
interest basis	compounding of interest
split interest	grace periods
interest payment date	minimum balance requirements
service charge rules	report options
holiday schedules	report formats

B. BENEFITS

1. The ability to process any type of time deposit.
2. Improved accuracy and control.
3. A data base large enough to provide for management information requests, and with capability for use as part of a future on-line inquiry system.
4. Ease of maintenance.

C. PROCESSING CYCLE

1. Daily- Activity capture, File update, Ledger reporting
2. Monthly-Workload data processing
3. Quarterly-Statement processing
4. Annually-Federal and state tax reporting.

D. REPORT OUTPUT

There are several basic reports (e.g., Daily Ledger, Statements, etc.) but each time deposit function and each user has the option of selecting additional or special reports.

Fig. 2-17 General System Narrative (Courtesy of Crocker Bank)

CROCKER NATIONAL BANK DATA PROCESSING AND SYSTEMS ADMINISTRATION	TITLE: OUTPUT/INPUT		Page PROJECT NO. SBP005
PROJECT: SAVINGS AND BONUS PASSBOOK		PREPARED BY: AN ANALYST	DATE PREPARED: 1/74
DEPARTMENT: SYSTEMS AND PLANNING		DEPARTMENT SUPERVISOR: A SUPERVISOR	APPROVED BY:

A. OUTPUT REPORTS

Master Error Report - Daily itemization of all type transactions, showing the data element(s) in error,with an indication of the type of error, such as check digit, non-numeric data, etc. The report informs the branch offices what errors were made on the OCR forms.
Daily Balance Listing - Daily itemization of Regular Savings and Bonus transactions, showing the status of interests and balances
BTR Recap and General Ledger Totals - Daily itemization of transactions showing number and type of accounts, current balances, various types of interests, etc. Also, month-to-date totals and General Ledger liability account totals. This report gives the user total debits and credits.
Exceptions Report - Daily itemization of all type transactions, showing the exceptions in the data element(s), with a message of the type of exception. This report informs the user of the unposted transactions and any exceptional transactions.
Year-to-date Transaction Report - Monthly itemization of all types transactions, showing year-to-date status.

B. USER SUPPLIED INPUT

The user is required to provide the following input in order to receive the above and additional reports.

Aside from the basic required input such as branch office name, number, phone, and account number, and the account holder's name, address, social security number, employer I.D. number, the input should include: the account number for interest credit. That is, account number credited when interest pay method is either credit to savings or credit to checking account.

C. PROCESSING CYCLE

1. Daily - File update, balance reporting, exception reporting, and General Ledger totals reporting.

2. Monthly - Interest payments on Bonus Passbook accounts, General Ledger income, and expense interface, year-to-date transaction reporting, statements, and maturity reporting.

3. Quarterly - Interest payments on regular savings, dormant account reporting, and statements.

4. Annually - Statements.

5. On Request - Management reporting.

Fig. 2-18 Output/Input (Courtesy of Crocker Bank)

description form (Figure 2-7), the input description form (Figure 2-12) and the work flowchart (Figure 2-9).

Make sure that you list the processing cycle, such as the daily, monthly, quarterly or "on request" reports separately on the output/input description form, Figure 2-18.

All through these steps keep in mind that you are a communicator. As much information as possible must be given to the user/management about the proposed system so that they can ascertain at the checkpoint/sign-off stage (at the end of this phase) whether you have designed a system according to the signed-off assignment and proposal, and most importantly, whether you have created a system that truly meets their requirements.

C/ CHECKPOINT/SIGN-OFF

The simpler you present your new system concept, the better chance you have of getting it approved and signed at this critical checkpoint. This is the last checkpoint before you can go ahead and develop the system. At this milestone both the user and you sign-off: The user formally agreeing not to add any other requirement; the systems man formally stating that the system will be delivered as promised.

Chapter
3

Solution Phase

This chapter describes in detail the method to be used in defining and documenting the activities of this structured phase to ensure the quality of the established information requirements.

During this phase the creativity and ingenuity of the systems analyst is put to test. It is in the detailed system design and perhaps even more so in the definitive program specifications task that most of the unforeseen difficulties of the proposed system become apparent.

Since it is extremely important that all systems decisions affecting the system structure be made *before* the actual programming (coding, testing, debugging) activity starts, make it a standard practice to define and document the program functional requirements during the program specifications task.

Because in this activity you must constantly refer to the output/report description (Figure 2-4) and the file description (Figure 2-5)—both in the previous phase—at the start of the solution phase have copies made of these forms. The former, as stated previously, describes the content and characteristics of the output/report; the latter states the physical attributes, content and characteristics of the files going outside the system and/or interfacing with other systems.

The solution phase consists of four definitive tasks: A/ detailed system design, B/ program specifications, C/ programming and D/ system test.

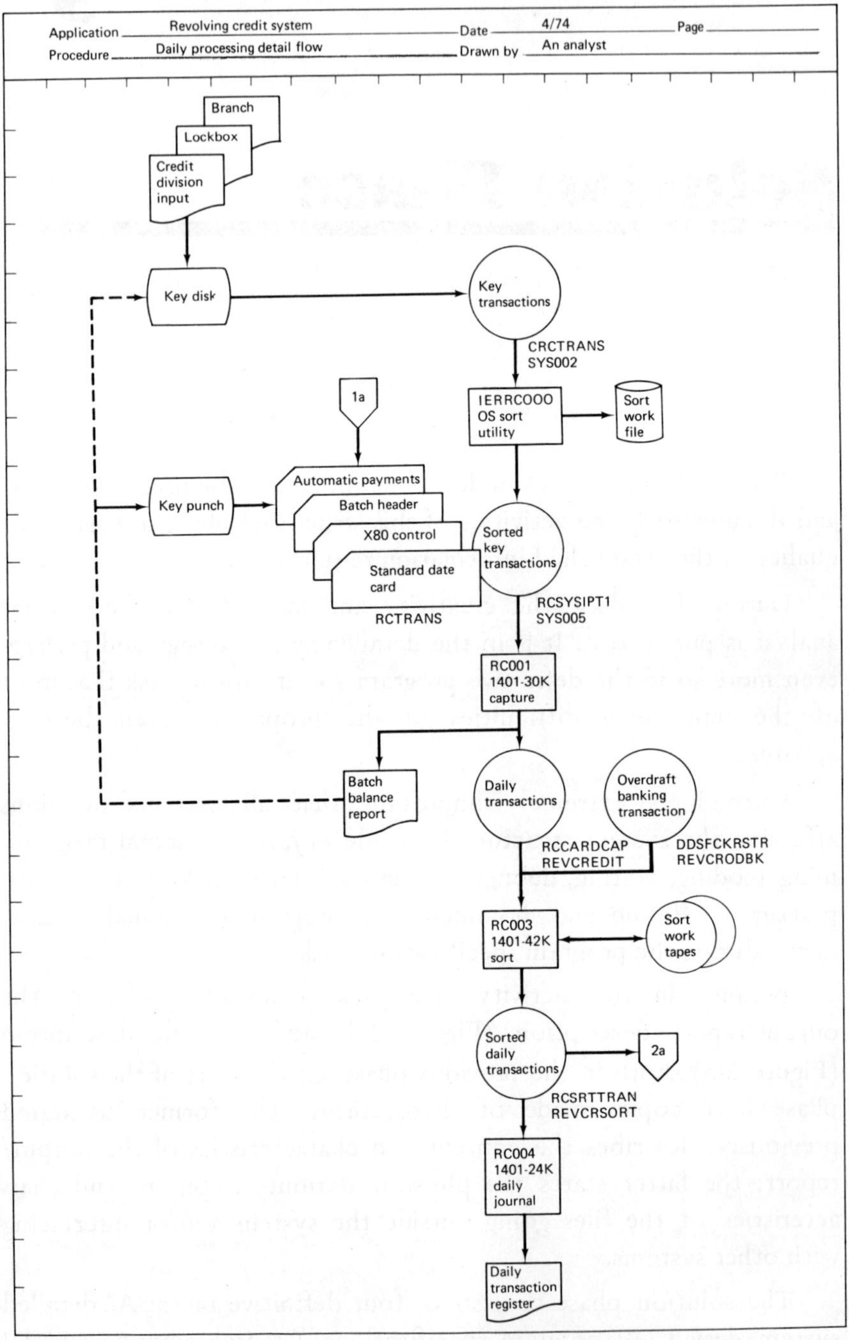

Fig. 3-1 Daily Processing Detail Flow

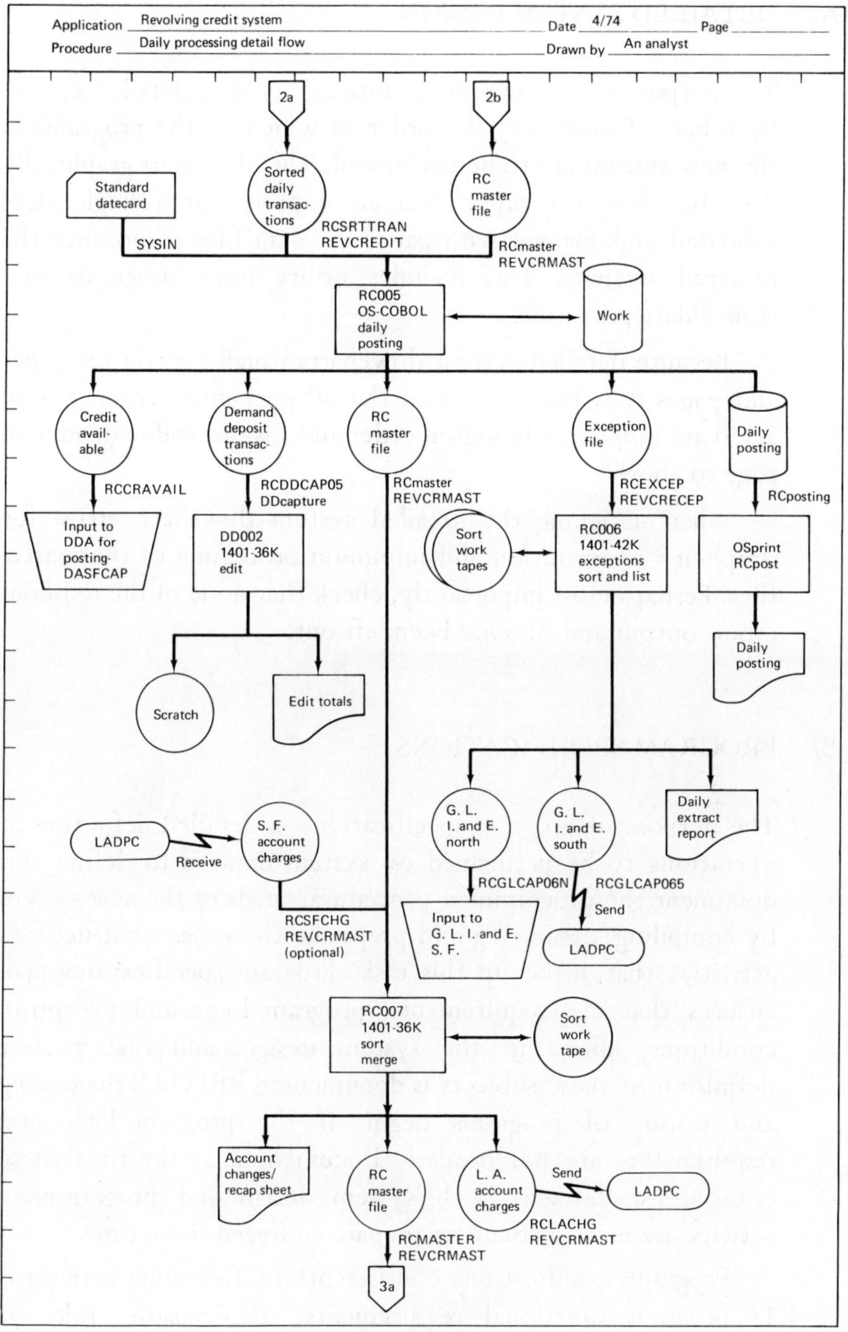

Fig. 3-1 (Continued)

A/ DETAILED SYSTEM DESIGN

The purpose of this task is to illustrate via the detailed system flowchart (Figure 3-1) the order in which all the programs in the new system are to be performed. This flowchart graphically describes how the input data are captured at a single step, validated and merged with particular data files to produce the required outputs. This includes utility print programs that consolidate print runs.

Because detailed system flowcharts usually go on for pages and pages it is important that the off-page connectors (Figure 2-15) are numbered in logical order and easy to follow from one page to another.

When designing the detailed system flowchart, strive for simplicity of operation and minimum processing of the master files. Perhaps most importantly, check that none of the required input, output and files has been left out.

B/ PROGRAM SPECIFICATIONS

The purpose of program specifications—a detailed definition of operations to be performed on system data—is to define and document the programmed procedural needs of the new system by compiling, organizing and preparing the series of structured activities that make up this task. Program specifications task ensures that the requirements, program logic and exception conditions adhere to the system design, and that a clear definition of these subjects is documented BEFORE the coding and testing of programs begin. If the program logic and requirements are not properly documented at the time these criteria are defined, both system design and programming activity are more difficult to prepare and need more time.

Program specifications consists of the following activities: 1/ program functional requirements, 2/ decision table, 3/ system input and 4/ data element definition.

1/ Program Functional Requirements (Figure 3-2)

To utilize this feature of the Program specifications, describe in detail for each program:

- Itemized, unique or nonstandard initialization requirements such as the date card and control card processing, table loading and so forth.
- Itemized, unique or nonstandard processing logic requirements relative to inter-data file relations such as descending keys, identification of primary and secondary files and so forth.
- Itemized data interpretation requirements, unique to each data file such as the sequence checking, record formatting, and selection of records for processing.
- Itemized application requirements unique to each program such as the computations employing data elements from more than one data file or table, updating, posting, and so forth.
- Itemized, unique or nonstandard I/O requirements for physical access of each data set, such as the sequential access of direct organized files, generic (partial key) access of indexed files, end-of-volume checkpoint rerun, and I/O error procedures.
- Itemized, unique or nonstandard termination requirements for processing, such as the run control totals, sequence error or other unrecoverable error termination. References to supporting documentation are to be included in all of these subsections.

2/ Decision Table (Figure 3-3)

Decision table provides an organized, concise means of communicating and documenting decision logic in the new system. It graphically describes cause and effect, and clarifies the conditions behind given actions. Decision table

CROCKER NATIONAL BANK DATA PROCESSING AND SYSTEMS ADMINISTRATION	TITLE: PROGRAM FUNCTIONAL REQUIREMENTS	Page PROJECT NO: SBP005
PROJECT: SAVINGS AND BONUS PASSBOOK	PREPARED BY: AN ANALYST	DATE PREPARED: 4/74
DEPARTMENT: SYSTEMS AND PLANNING	DEPARTMENT SUPERVISOR: A SUPERVISOR	APPROVED BY:

SBP04 - BRANCH TABLE MAINTENANCE

Initialization Requirements

Access information passed to program in PARM field of EXEC card. Move date field CD-DATE of CTL-CD-REC to Branch Table List heading date field Date-P. Move Create/Update indicator in 'PARM ________' to FUNCTION field in working storage.

Logic Requirements

Primary file for files processing control is the Branch Table file TABLE. Key for files processing control is BRANCH-NO. Any delete transaction on Update Data cards file CARDBR must match on key to TABLB file record TABLB-REC. Any add transaction must not match on key to TABLE-REC. Any change transaction must match on key to TABLE-REC. (Ref: DT, 4.4.3, page 3)

Data Interpretation Requirements

Validate CTL-CD-REC fields according to system editing rules. (Ref:IEF, 4.5.8, pagel) Test value of Function for 'C'. When test is true, set value of BYPASS-READ-SWITCH equal to '1' (ON). When falue of BYPASS-READ-SWITCH is not equal to '1' (OFF), sequence check TABLB file on BRANCH-NO key, ascending. Sequence check CARDBR file on BRANCH-NO key, ascending.
Validate field content of CARDBR file record CARDBR-REC according to system editing rules. (Ref: IEF, 4.5.8, page 1) CARDBR file can contain up to 3 records (identified by values 1, 2, and 3 in CARD-NO field) for a branch number key. Construct a logical record in working storage for each key from 1, 2, or 3 records, as applicable, and apply working storage record to TABLB file.

Application Requirements

Apply delete transactions (TRAN-CODE D).
Delete (bypass writing) TABLB-REC when value of the number of master records (MSTR-REC) field of TABLB-REC is equal to zero.
Set the value of DELETE-INDfield of TABLB-REC equal to 'D' when MSTRA-RECS is not equal to zero, and the number of open accounts OPEN-ACCTS field of TABLB-REC is equal to zero. Reject CARDBR-REC when OPEN-ACCTS is not equal to zero.

Apply add transactions (TRAN-CODE A).
When adding new TABLB-REC set value of DEPOSITS, WITHDRAWALS, INTEREST PAID, and BALANCE fields in TABLB-REC equal to zero.

APPLY change transactions (TRAN-CODE C).
Replace value of TABLB-REC fields with value of corresponding non-blank CARDBR-REC fields, then test the DELETE-IND field for 'D' and test MSTR-RECS for zeros. When both conditions are true, delete TABLE-REC. (continued on next page)

Fig. 3-2 Program Functional Requirements (Courtesy of Crocker Bank)

CROCKER NATIONAL BANK DATA PROCESSING AND SYSTEMS ADMINISTRATION	PROGRAM FUNCTIONAL REQUIREMENTS (cont'd) TITLE	Page PROJECT NO SBP005
PROJECT SAVINGS AND BONUS PASSBOOK	PREPARED BY AN ANALYST	DATE PREPARED 4/74
DEPARTMENT SYSTEMS AND PLANNING	DEPARTMENT SUPERVISOR A SUPERVISOR	APPROVED BY

Input/Output Requirements

When BYPASS-READ-SWITCH is equal to 1, bypass reading TABLB file.
When DELETE-IND value is equal to'V', list TABLB-REC.
When equal to 'D', do not list.

Termination Requirements

When sequence error on TABLB file occurs, display prior table key and TABLB-REC key, and terminate processing.
When sequence error on CARDBR file occurs, display prior card key and CARDBR-REC key, and terminate processor.
When ADD transaction key matches TABLB-REC key, display CARDBR-REC key, and message ADD-ALREADY ON TABLE, and terminate processing.
When CHANGE transaction key does not match TABLB-REC key, display CARDBR-REC key, and message CHANGE-NOT ON TABLE, and terminate processing.
When DELETE transaction key does not match TABLB-REC key, display CARDBR-REC key, and message DELETE-NOT ON TABLE, and terminate processing.

Fig. 3-2 (Continued)

comprises *conditions* and *rules*. The first is a statement of *conditions* that may be present within the system; the second statement defines the *actions* that must be taken based on the presence of certain conditions.

There are two major types of decision tables. The first one is the LIMITED ENTRY TABLE (Figure 3-3a), which expresses each condition as a question in the condition stub. In the rules, the condition value is expressed as Y (yes), N (no) and blank or dash, meaning indifference. In the action stub, actions are specified by X (execute), and blank or dash (ignore). The limited decision table may be used by analysts who are unfamiliar with this effective tool. The second type of decision table is the EXTENDED ENTRY TABLE (Figure 3-3b) in which only part of the condition is stated in the condition stub, and only part of the action is stated in the action stub.

The remainder, i.e., the variable either in the condition or the action is stated in the entry itself. Whether using limited or extended decision table, "legends" at the

		Decision Rule 1	Decision Rule 2	Decision Rule 3	Decision Rule 4
Condition Stubs:	Credit satisfactory	Y	N	N	N
	Pay history favorable	–	Y	N	N
	Clearance obtained	–	–	Y	N
Action Stubs:	Approve order	X	X	X	–
	Return order to dept.	–	–	–	X

Legend:
Decision Rule 1. If customer's credit is satisfactory, approve the order.
Decision Rule 2. If customer's credit is not satisfactory, but his pay history favorable, approve the order.
Decision Rule 3. If customer's credit is not satisfactory, and his pay history is not favorable, but clearance obtained, approve the order.
Decision Rule 4. If customer's credit is not satisfactory, and his pay history is not favorable, and clearance is not obtained, return order to the Order Processing department.

Fig. 3-3a Limited Entry Table (Decision Table)

CROCKER NATIONAL BANK
DATA PROCESSING AND SYSTEMS ADMINISTRATION

EXTENDED ENTRY TABLE

Page

PROJECT: SAVINGS AND BONUS PASSBOOK	PREPARED BY: AN ANALYST	PROJECT NO: SBP005
DEPARTMENT: SYSTEMS AND PLANNING	DEPARTMENT SUPERVISOR: A SUPERVISOR	DATE PREPARED: 2/74
TABLE TITLE: DELETE FUNCTION	TABLE NAME: T012	APPROVED BY:

LINE	CONDITION/ACTION RULE ➡	1	2	3	4	5
C01	TRANSACTION CODE EQUAL TO 'D'	N	Y	Y	Y	Y
C02	BRANCH NUMBER OF TRANSACTION EQUAL TO BRANCH NUMBER OF TABLE FILE		N			
C03	TABLE FILE 'MSTR-RECS' FIELD VALUE EQUAL TO ZERO				N	Y
C04	TABLE FILE 'OPEN-ACCTS' FIELD VALUE EQUAL TO ZERO			N	Y	Y
A01	REJECT TRANSACTION			X		
A02	SET VALUE OF TABLE FILE 'DELETE-IND' FIELD EQUAL TO 'D'				X	
A03	WRITE OUTPUT TABLE FILE RECORD				X	
A04	READ TRANSACTION FILE			X	X	X
A05	READ TABLE FILE				X	X
A06	GO TO T013 (ADD FUNCTION)	X				
A07	GO TO T002 (MAINLINE)			X	X	X
A08	TERMINATE PROCESSING (LOGIC ERROR)		X			
	GO TO-- F (FUNCTION); R (RULE, SAME TABLE); T (TABLE)					

Fig. 3-3b Extended Entry Table (Decision Table) (Courtesy of Crocker Bank)

bottom of the table should explain the various decision rules. Prepare a decision table for each program, and make sure that you define the input data elements and their effect on the master file data elements. In addition, ensure that you define the input editing and validation rules, as well as the major processing functions to be performed on specific input transactions.

3/ System Input

Before preparing the system input, which provides the necessary documented information on input for programming, you must decide which method (punched card, key-to-tape/disk, on-line terminal) is best for producing information in a machine-readable form for the new system. Once the medium is established, you can develop the input accordingly.

The input data description (Figure 3-4) describes the sources, volumes and procedures associated with the new or revised source documents that contain information required as input to the new system. The input data description is also used to identify new or revised manual functions, and to determine personnel requirements. This description should be accompanied by a rough layout of the new or revised source documents.

When designing source documents, bear in mind that the system must be designed within the operating personnel capabilities.

When designing the forms, ensure that they are easy to read and easy to complete. Have adequate instructions preprinted on the forms, and use precise wording. Avoid incomplete abbreviations; they can be misinterpreted. Exercise caution. Input accuracy and simplicity are mandatory. Economy, though important, may be far outweighed by subsequent system problems.

Much of the data needed for system input (such as file description, Figure 2-5) have been developed and docu-

CROCKER NATIONAL BANK
DATA PROCESSING AND SYSTEMS ADMINISTRATION

INPUT DATA DESCRIPTION

Page

PROJECT	PREPARED BY	PROJECT NO
		TDS1234
TIME DEPOSIT SYSTEM	AN ANALYST	DATE PREPARED: 4/74
DEPARTMENT: BANK OPERATIONS	DEPARTMENT SUPERVISOR: A SUPERVISOR	APPROVED BY

DOCUMENT NO	COPY ATTACHED	FORM TITLE	FORM NO
Do-1	☒YES ☐NO	NAME AND ADDRESS NOTICE	Do-1

FORM PURPOSE

To enter name and address information for savings accounts into the ADA Name and Address System

PRESENT INPUT FORM

Preprinted form

PROPOSED INPUT FORM

CRT Display Terminal

PROCESS AND EDIT STEPS	PERFORMED BY
Enter data using CRT Display Terminal & verify data on tape	Terminal Clerk
Retain one copy in branch files; forward one copy to DPC	Op. Supervisor
Submit tape for computer processing.	Control Clerk, DPC
Return Exceptions listings and Updates to branches	Control Clk, DPC

CONTROL	VOLUME				
	PERIOD	FREQ.	MIN.	AVG.	MAX.
Program RSA01 - Exceptions Listing	Jan-Feb	Weekly	5	150	200

COMMENTS

Fig. 3-4 Input Data Description (Courtesy of Crocker Bank)

CROCKER NATIONAL BANK
DATA PROCESSING AND SYSTEMS ADMINISTRATION

TITLE		Page PROJECT NO SBP005
PROJECT SAVINGS AND BONUS PASSBOOK	PREPARED BY AN ANALYST	DATE PREPARED 4/74
DEPARTMENT SYSTEMS AND PLANNING	DEPARTMENT SUPERVISOR A SUPERVISOR	APPROVED BY

DATA ELEMENT DEFINITION

Descriptive Name TYPE-CODE

Field Name TYPE

FORMAT
Input 99
File 99
Output 99

Purpose & Description TWO DIGIT CODE INDICATING TYPE OF ACCOUNT.

Method of Updating Extension INCORRECT TYPE CODES ARE REPLACED WITH CORRECTION INPUT DURING MASTER FILE MAINTENANCE.

Source Documents	Input Records	Output Records	Files
D01, D02,	MSTRA-REC	MSTRA-REC	SB.MSTRA
D12, D13,	TRANS-REC	TRANS-REC	SB.OUTPGL
D17, D18,	REVLD-REC	TRLBR-REC	SB.REPTFDIC
D03, D05,	DTRS-REC	NATR-REC	SB.CARDBAL
D15, D16,	ELRP-REC	LRCP-REC	SB.CNTLD1
D19, D21,	INEXP-REC	PNSE-REC	SB.CNTLD2
D11, D06	IEHDR-REC	XRPT-REC	SB.TABLB

NOTE: If additional entries are required for source documents, input records, etc., use General Purpose Form No. 32-200 as a second page. (Continued on next page)

Fig. 3-5 Data Element Definition (Courtesy of Crocker Bank)

CROCKER NATIONAL BANK DATA PROCESSING AND SYSTEMS ADMINISTRATION	TITLE	Page PROJECT NO SBP005
PROJECT SAVINGS AND BONUS PASSBOOK	PREPARED BY AN ANALYST	DATE PREPARED 4/74
DEPARTMENT SYSTEMS AND PLANNING	DEPARTMENT SUPERVISOR A SUPERVISOR	APPROVED BY

DATA ELEMENT DEFINITION (Cont'd)

Code Values	Description
21	REGULAR SAVINGS
22	THREE MONTH BONUS PASSBOOK
24	ONE YEAR BONUS PASSBOOK
25	TWO YEARS BONUS PASSBOOK
20	TRUST DEPARTMENT

Fig. 3-5 (Continued)

mented in the development phase and need only to be duplicated. Two other sources for system input are the decision table (Figure 3-3a or 3-3b), and the data element definition form, which is the next and last step in program specifications.

4/ Data Element Definition (Figure 3-5)

The data element definition form defines and documents the system data structure by compiling, organizing and preparing the data elements. It combines the system coding structure, (i.e., it states the codes within the system and defines their descriptive titles), the data element dictionary and the data element cross-reference chart. To keep it uncomplicated, prepare one sheet per data element.

C/ PROGRAMMING

The purpose of the programming task is to code, test and debug each program within the system. The actual technique of coding, testing and debugging is a book subject by itself, and therefore does not belong in this book. What is discussed here is how this activity can be simpler yet more efficiently performed by documenting several aspects of this task with the aid of the detailed system flowchart (Figure 3-1), the decision table (Figure 3-3) and the data element definition (Figure 3-5). Having these data, the programmer (under the systems analyst's supervision) should write a program description (Figure 3-6) for every one of the programs in the system.

The program description should have a one-paragraph abstract of what the program accomplishes, followed by detailed description of the functions of the particular program. Each paragraph of the detailed narrative should start with a section or paragraph name before describing its function in a concise and clear manner. Next step should be the test procedures (Figure 3-7), which narrate how each program is

Page

CROCKER NATIONAL BANK DATA PROCESSING AND SYSTEMS ADMINISTRATION	TITLE: PROGRAM DESCRIPTION	PROJECT NO: TDC0201D
PROJECT: ACTIVITY CAPTURE	PREPARED BY: A. PROGRAMMER	DATE PREPARED: 4/74
DEPARTMENT: BANK OPERATIONS	DEPARTMENT SUPERVISOR: A. SUPERVISOR	APPROVED BY:

ABSTRACT

TDC0201D merges Daily Activity input with corrected Activity input and validates each record. Exceptions are listed together with error codes. Valid records are reformatted and sorted.

DETAIL

A10-HSK reads the standard date card from SYSIN, initializes Working-Storage entries and opens all files.

A-20-READ-BRANCH reads BRANCH-TABLE and stores the entries in ascending sequence in BR-TBL. (The format and other details of BRANCH-TABLE are described in the attached Table Description).

B10-MAIN executes C10-MRG-ACTIVITY to select an activity record from ACTIVITY-DLY and ACTIVITY-HLD. Each activity record is validated according to the attached Input Edit form.

E10-WRITE-EXCP lists invalid transaction. The listing shows the transaction record together with the appropriate error indication.

B20-EXPAND expands valid transactions to contain DATE-TODAY and executes F10-WRITE-ACTIVITY to write ACTIVITY-CAPTD.

B30-END closes all files, displays the transaction count and number of invalid transactions counted by C10-MRG-ACTIVITY and E10-WRITE-EXCP respectively, and executes STOP RUN.

The utility Sort follows TDC0201S and sorts ACTIVITY-CAPTD on ACCT-NO (major), TRANS-DATE and TRANS-CODE (minor), producing ACTIVITY-SRTD. A listing of the sort control cards is attached.

Fig. 3-6 Program Description (Courtesy of Crocker Bank)

CROCKER NATIONAL BANK
DATA PROCESSING AND SYSTEMS ADMINISTRATION

TITLE: TEST PROCEDURE		PAGE / PROJECT NO: TDC0201D
PROJECT: ACTIVITY CAPTURE	PREPARED BY: A. PROGRAMMER	DATE PREPARED: 4/74
DEPARTMENT: BANK OPERATIONS	DEPARTMENT SUPERVISOR: A. SUPERVISOR	APPROVED BY:

TEST 1

INPUT: Date Card with column 1 other than X punch (-). No other input required.

TEST PLAN: TDC0201D should display on console "INVALID DATE CARD-RERUN JOB" and terminate.

RESULTS: none

TEST 2

INPUT: Date Card punched per attachment 1
ACTIVITY-DLY as listed on attachment 2
ACTIVITY-HELD as listed on attachment 3 (use IEBGENER with control deck as listed on attachment 4)
BRANCH-TABLE as listed on attachment 5 (use IEBGENER with control deck as listed on attachment 6)

TEST PLAN: Set up test deck per JCL listing on attachment 7. This will list the exception report followed by ACTIVITY-CAPTD on SYSOUT.

RESULTS: EXCEPTIONS as shown on attachment 8
ACTIVITY-CAPTD as shown on attachment 9

Fig. 3-7 Test Procedure (Courtesy of Crocker Bank)

validated in terms of input data, internal events and results. Simple programs may get by with one test run, but complex programs usually need two or more test runs for satisfactory verification.

Each documented test should include the following headings: Input, Test Plan, and Results. Under input list all the test data being used; under test plan state the events that should occur and under results describe the expected results. It is a good practice to record the test results of each program on the test result log (Figure 3-8). It provides an accurate history of each program test and its results, together with the program listing, might be helpful at a later date to the person who has to trace the evolution of a particular program for modification. Final step in the programming task is debugging. There are many debugging techniques including debugging "packets," depending on the hardware, program language being used and the facility. The two major types of debugging are the source level debugging and the machine level debugging. The former identifies and corrects errors in the source program; the latter uses procedure maps and memory dumps to locate errors and relate them with source statements. There is also what is called "selective testing" (Figure 3-9) in which any number of debugging packets may be used to modify or superimpose a flow of control different from the normal control flow and perform input-output processing.

D/ SYSTEM TEST

The purpose of the system test task is to ensure that all the system's ingredients work together well to effect a system as designed by the analyst. To ensure that all the major activities are included in the system test (computer programs, data processing operations and user activities) document all major activities in relation to responsibility, schedule and approval-sign off on the system test checklist (Figure 3-10). Moreover, the system test should not start until and unless all components on the checklist have been performed.

CROCKER NATIONAL BANK
DATA PROCESSING AND SYSTEMS ADMINISTRATION

RESULTS LOG

[X] TEST PROGRAM   [ ] IMPLEMENTATION

Page
PROJECT NO: TDS1234

PROJECT	PREPARED BY	DATE PREPARED
TIME DEPOSIT SYSTEM	A PROGRAMMER	1/74
DEPARTMENT	DEPARTMENT SUPERVISOR	APPROVED BY
BANK OPERATIONS	A SUPERVISOR	

DATE OF TEST	TIME OF TEST	TEST	OPERATOR RESPONSIBLE	programmer RESPONSIBLE	EA ✓	RESULTS
1/2	10 a.m.	T2		A. Jones		S.R. 537 Test Results O.K.
1/6	3 a.m.	T2		B. Smith		S.R. 935 Branch Table Overflow
1/27	2 p.m.	T2		C. White		S.R. 178 Test Results O.K.

Fig. 3-8 Results Log—Test Program (Courtesy of Crocker Bank)

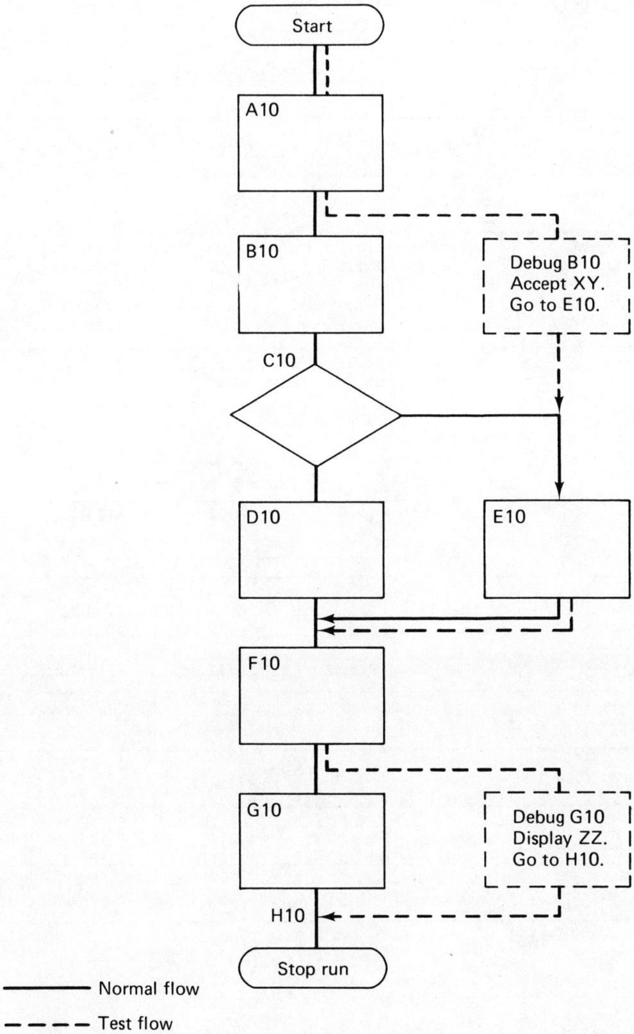

Fig. 3-9 Selective Testing–Using Debugging Packets

CROCKER NATIONAL BANK
DATA PROCESSING AND SYSTEMS ADMINISTRATION

SYSTEM TEST CHECKLIST

Page

PROJECT	PREPARED BY	PROJECT NO TDS1234
TIME DEPOSIT SYSTEM	AN ANALYST	DATE PREPARED 4/74
DEPARTMENT BANK OPERATIONS	DEPARTMENT SUPERVISOR A SUPERVISOR	APPROVED BY

TEST:	SCHEDULED START DATE	ACTUAL START DATE	SCHEDULED COMPLETION	ACTUAL COMPLETION	PROJECT LEADER
YEAR-END PROCESSING	4/12/74	4/12/74	4/13/74	4/13/74	B. Smith

DESCRIPTION OF ACTIVITY	INDIVIDUAL RESPONSIBLE	DEPARTMENT RESPONSIBLE	START DATE	COMPLETION DATE	APPROVED BY
1. TEST DESCRIPTIONS	T. Stark	Auto.Syst.	3/28/74	4/02/74	
2. TEST PROCEDURES	B. Smith	Auto.Syst.	2/10/74	2/28/74	
USER DEPARTMENT	T. Tom	Proc. Div.	2/10/74	2/28/74	
DATA PROCESSING OPERATIONS	B. Smith	Auto.Syst.	2/10/74	2/28/74	
SYSTEMS & PROGRAMMING	B. Smith	Auto.Syst.	2/10/74	2/28/74	
DATA PROCESSING CONTROLS	B. Smith	Auto·Syst.	2/10/74	2/28/74	
3. FORMS	D. Wont	Ba Proc/Frms	12/29/73	2/9/74	
INPUT TEST DATA	D. Wong	Ba Proc/Frms	12/29/73	2/9/74	
FILE CREATION & MAINTENANCE					
4. CONTROLS					
USER DEPARTMENT	T. Tom	Proc.Div.	4/12/74	4/25/74	
DATA PROCESSING CONTROLS	B. Smith	Auto.Syst.	4/12/74	4/16/74	
5. EQUIPMENT	M. Washington	DPC	4/12/74	4/13/74	
AVAILABILITY					
EMERGENCY SERVICE					
6. MANPOWER					
USER DEPARTMENT	I. Dale	Pilot Bch	4/01/74	4/28/74	
DATA PROCESSING OPERATIONS	M.Wash.	DPC	4/12/74	4/13/74	
SYSTEMS & PROGRAMMING					
DATA PROCESSING CONTROLS	J. Martin	CkProc&Acct	4/12/74	4/13/74	
7. SUPPLIES					
AVAILABILITY					
8. DATA					
CREATED & VERIFIED (User)	W. Brady	Proc.Div.	4/01/74	4/28/74	
CONTROLLED	B. Smith	Auto. Syst.	4/12/74	4/16/74	
9. FILES					
CREATED & VERIFIED (User)	T. Tom	Proc.Div.	4/01/74	4/28/74	
CONTROLLED	B. Smith	Auto. Syst.	4/12/74	4/16/74	
10. MISCELLANEOUS					

COMMENTS

☐ CHECK IF MULTI-PURPOSE CHECKLIST FORM ALSO USED

Fig. 3-10 System Test Checklist (Courtesy of Crocker Bank)

CROCKER NATIONAL BANK
DATA PROCESSING AND SYSTEMS ADMINISTRATION

TEST DESCRIPTION

Page

PROJECT NO.
TDS1234

PROJECT	PREPARED BY	DATE PREPARED
TIME DEPOSIT SYSTEM	AN ANALYST	3/74
DEPARTMENT	**DEPARTMENT SUPERVISOR**	**APPROVED BY**
BANK OPERATIONS	A SUPERVISOR	

TEST:	DATE SCHEDULED	DATE COMPLETED	OPERATOR RESPONSIBLE	EQUIPMENT TO BE USED	ANALYST RESPONSIBLE
YEAR-END PROCESSING	4/12/74	4/13/74	A. Adams	360/65	B. Smith

TEST DESCRIPTION

This system test will verify that all year-end processing requirements have been satisfied. The entire system will be run with a test date of 12/29/73. The History and Ledger files will be converted for two branches and actual input data for that date (activity) will be used.

The following requirements will be checked:

a. I.R.S. Report

b. Y-T-D Interest cleared

c. Purge Ledger, drop closed accounts

d. Purge History

e. Update "account beginning balance" fields

f. Clear debit count

g. Prepare statements.

The I.R.S. Reports will be verified according to the procedures set forth in the Bank Manual Sec. 856-3.

PROGRAM NO.	PROGRAMMER	PROGRAM NO.	PROGRAMMER	PROGRAM NO.	PROGRAMMER	PROGRAM NO.	PROGRAMMER
AAEX	Walker	AAUX	Colter				
AADX	Jonas	AAVX	Mason				
AAIX	Warren	AAWX	Carter				
AAPX	Jackson	AACX01	Watson				
AARX	Clay	AACX20	Pearson				
AASX	Smith						

COMMENTS

The TDS system pack PR0009 contains all the table and master data sets for this test.

Fig. 3-11 Test Description (Courtesy of Crocker Bank)

CROCKER NATIONAL BANK
DATA PROCESSING AND SYSTEMS ADMINISTRATION

RESULTS LOG

☒ SYSTEM TEST ☐ IMPLEMENTATION

Page

PROJECT NO: TDS1234

PROJECT: TIME DEPOSIT SYSTEM

PREPARED BY: AN ANALYST

DATE PREPARED: 3/74

DEPARTMENT: BANK OPERATIONS

DEPARTMENT SUPERVISOR: A SUPERVISOR

APPROVED BY:

DATE OF TEST	TIME OF TEST	TEST	OPERATOR RESPONSIBLE	ANALYST RESPONSIBLE	EA ✓	RESULTS
3/12	9 a.m.	YEAR-END AAEX	R.Anders	B. Smith		SCC-522
3/12	10 a.m.	YEAR-END AAEX	J. Bolt	B. Smith		Exception Report incomplete
3/13	9 a.m.	YEAR-END AAPX	S. Deers	B. Smith		Y-T-D Interest not cleared on Ledger
3/13	9 a.m.	YEAR-END AAUX	P.Jones	B. Smith		Overflow on statements being lost

Fig. 3-12 Results Log—System Tests (Courtesy of Crocker Bank)

Subsequently, when the test is accomplished, i.e., when all data have passed a preestablished number of error-free cycles, document or rather describe each test on the test description form (Figure 3-11).

To ensure that the system meets user's satisfaction, a verification of system test results should be performed by the user. Finally, a system test results Log form should be filled out. System test results log (Figure 3-12) provides not only a chronological record of all the test runs and the results, but also a tool for systems and data processing management as well as user management to follow the system tests and evaluate the personnel responsible for the test. In addition, the system test results log supplies a systematic approach to solution of problems discovered during the tests.

Chapter
4

Installation Phase

This chapter describes the methodical steps that must be performed *prior* to initiating the installation of the system; the training of the user, including the very important but often overlooked user's manual and the installation of the new or revised system.

In this phase the emphasis is on the meticulous care with which the planning of the installation of the system has to be accomplished. A carefully prepared and carried out conversion/installation planning activity is essential for the successful implementation of the system project. However, if the system testing has been comprehensive and well documented, and if the systems man does a thorough job in the planning stage, the actual installation should be a comparatively easy job.

The installation phase comprises three tasks: A/ conversion/ installation planning, B/ user training/user's manual and C/ installation.

A/ CONVERSION/INSTALLATION PLANNING

The purpose of this task is to define and document the definitive steps by which the transformation is to be performed so that there will be minimal side effects because of the change.

To ensure a smooth and orderly transition to the new or modified system, indicate the person(s) responsible for the various steps in this task and establish the schedule for the designated activities listed on the conversion/installation planning checklist (Figure 4-1). This form should be approved/signed-off by you as well as a representative of each affected department.

Next, describe briefly and concisely the procedure(s) for each required task on the conversion/installation plan form (Figure 4-2). The numbers within the text on this form correspond to the sections on the checklist that need to be elaborated.

To augment the checklist and the narrative plan, or rather to present graphically to the user/management the schedule of the major events within the installation phase, prepare the conversion installation planning chart (Figure 4-3).

Finally, on the manpower and equipment plan form (Figure 4-4)–based on quantitative analyses and estimates–project the resources needed to install a new or revised system. The work measurement department/group (if there is one) can make this task much easier for you by providing data of the manpower by skill level, and the equipment hours needed.

The following are the minimum activities required during the conversion/installation planning task: 1/ create or modify files, 2/ write conversion programs (if applicable), 3/ prepare and introduce new forms, and phase out old forms, 4/ write a runbook (for the computer operations personnel) and keying instructions (for the keypunch or key-to-disk/tape operators), 5/ write procedures, i.e., supervise the writing of procedures by a technical writer for all personnel involved in the processing of the scheduled new or revised system, such as input/output clerks and data control clerks–distribution instructions should also be written for the document control librarian, 6/ arrange and monitor parallel processing operation of the current and the new system, or pilot operation.

1/ Create/Modify Files

When creating new files, define and document the source of each data element and the method you use to establish

CROCKER
NATIONAL BANK
DATA PROCESSING AND
SYSTEMS ADMINISTRATION

CONVERSION PLANNING CHECKLIST

Page

PROJECT NO: RCC064

PROJECT	PREPARED BY	DATE PREPARED
Retail Credit Card Processing	An ANALYST	4/6/74
DEPARTMENT	**DEPARTMENT SUPERVISOR**	**APPROVED BY**
RETAIL CREDIT CARD	A SUPERVISOR	

TASK DESCRIPTION	INDIVIDUAL RESPONSIBLE	DEPARTMENT RESPONSIBLE	START DATE	COMPLETION DATE	APPROVED BY
1. CONVERSION PLAN	C. Hays	Auto Sys	4/5/74	4/9/74	
2. CONVERSION CHART	C. Hays	Auto Sys	4/5/74	4/9/74	
3. FILE DATA ACQUISITION PLAN	N/A				
4. MANPOWER & EQUIPMENT PLAN	C. Hays	Auto Sys	4/5/74	4/9/74	
5. USER TRAINING - CHECKLIST	Parsons	RCCD	4/5/74	4/9/74	
6. USER TRAINING - PLAN	Parsons	RCCD	4/5/74	4/9/74	
7. NEW FORMS	N/A				
PRINTING SPECIFICATIONS					
ORDERING					
CONVERSION					
8. CONVERSION PROGRAMS	C. Turner	Auto Sys	4/5/74	4/6/74	
DEFINATION					
SPECIFICATIONS					
PROGRAMMING/TESTING					
9. FILES	N/A				
PURIFICATION OF RECORDS					
DATA CONVERSION					
MAINTENANCE					
CONTROL					
10. INTERFACE WITH OTHER SYST.	N/A				
FILES					
DATA					
PROGRAMS					
11. EDP EQUIPMENT	N/A				
12. EDP PACKAGES/SERVICE	Parsons	RCCD	Complete		
Data Copy, Inc.					
13. SUPPLIES	A. Holmstedt	Auto Systems	4/5/74	4/9/74	
Address Labels					
Mailers					
14. Production JCL	F. Hambrick	TM&T	Started	4/9/74	
Operations Documentation	Holmstedt		4/14/74	4/23/74	

☐ CHECK IF MULTI-PURPOSE CHECKLIST FORM ALSO USED

Fig. 4-1 Conversion Planning Checklist (Courtesy of Crocker Bank)

CROCKER NATIONAL BANK DATA PROCESSING AND SYSTEMS ADMINISTRATION	TITLE: CONVERSION PLAN		Page PROJECT NO: RCC064
PROJECT: RETAIL CREDIT CARD PROCESSING		PREPARED BY: AN ANALYST	DATE PREPARED: 4/6/74
DEPARTMENT: RETAIL CREDIT CARD		DEPARTMENT SUPERVISOR: A SUPERVISOR	APPROVED BY:

Reference
Checklist

#8 Conversion Programs

Modifications need to be made to the utility program "Superu" in order to convert present non-monetary input into the new system format for parallel processing.

#12 EDP Service

B. Parsons of Retail Credit Card Department has informed Data Copy, Inc., to discontinue their services as of 4/30/74.

#13 Supplies

Address labels need to be ordered, and the supply of mailing inserts transferred from RCC Department to the data center.

#14 Production JCL

All JCL (Job control Language) is complete, except for one program which should be ready by 4/9/74.

Fig. 4-2 Conversion Plan (Courtesy of Crocker Bank)

CROCKER NATIONAL BANK
DATA PROCESSING AND SYSTEMS ADMINISTRATION

GENERAL PURPOSE CHART

Page

TITLE	PROJECT NO
CONVERSION PLANNING	RCC064

PROJECT	PREPARED BY	DATE PREPARED
RETAIL CREDIT CARD PROCESSING	AN ANALYST	3/6/74
DEPARTMENT	**DEPARTMENT SUPERVISOR**	**APPROVED BY**
RETAIL CREDIT CARD	A SUPERVISOR	

ACTIVITY	REF	MARCH 4	11	18	25	APRIL 1	8	15	22		
Conversion Checklist & Plan	1,2						—				
Manpower & Equipment Plan	4						—				
User Training/User's Manual							—	—			
New Forms		—	—	—	—	—	—				
Supplies							—	—			
Production JCL								—	—		
Operations Documentation								—	—		
Programming Effort											
Program Testing		—	—	—	—	—					
System Testing							—				
Parallel Processing								—			
Installation									→		

Fig. 4-3 General Purpose Chart—Conversion Planning (Courtesy of Crocker Bank)

CROCKER NATIONAL BANK
DATA PROCESSING AND SYSTEMS ADMINISTRATION

MANPOWER AND EQUIPMENT PLAN

Page

PROJECT		PREPARED BY	PROJECT NO.
RETAIL CREDIT CARD	PROCESSING	AN ANALYST	RCC064
DEPARTMENT		DEPARTMENT SUPERVISOR	DATE PREPARED
RETAIL CREDIT CARD		A SUPERVISOR	4/6/74
			APPROVED BY

RESOURCE	SKILL LEVEL	HOURS REQ'D.	SCHEDULE					
			4/5 - 4/16	4/19 - 4/23	4/26 - 4/30			
1. USER DEPARTMENT								
Retail Credit Card	Supervisor		12.0	12.0	4.0			
	Clerk		4.0	2.0	2.0			
2. DATA CONTROL								
N/A								
3. KEYPUNCH								
N/A								
4. OUTSIDE SERVICES								
N/A								
5. EAM EQUIPMENT								
N/A								
6. COMPUTER								
360 - 65		3.9	2.4	1.5				
7. OTHER								
O.C.R. - 915		.7	.5	.2				
COM		.1	.1					

COMMENTS

Fig. 4-4 Manpower and Equipment Plan (Courtesy of Crocker Bank)

these data elements in the files. Also, describe the procedures performed to ensure data integrity.

When converting existing data files into the new or revised system formats, future processing problems may be avoided or minimized by ensuring that the data carried forward from the old to the new system are valid as defined in the new system limits, ranges and relationships, Data integrity may be accomplished by writing a program that will convert master file data from the old format to the new. This program places on the new file only fields of historical nature that are not input data elements of the new system (for example, year-to-date accumulators). All other fields in the new master file are initialized as zeros or spaces, as applicable. The balance of the old master file fields are written out onto the other files in the formats acceptable as input to the new system. These files are then passed through the new system input edit program and, ultimately, through the new system master file update program to produce the new system master file. There are several methods to ensure data integrity, but the above approach is perhaps the simplest and most effective one.

2/ Write Conversion Programs

If a present system is being modified, conversion programs have to be written to convert certain files, tables, etc., into the revised system format. For example, if an existing table needs to be read and reformated to disk, a conversion program will accomplish such requirement.

3/ Prepare and Introduce New Forms and Phase Out Old Forms

Since this activity is fully covered in Part 2 of this book, this is just a reminder to design the new forms with simplicity and legibility in mind. If the new forms are easy to understand, uncrowded and simple to fill out, they will be readily accepted by all personnel, making your job that much easier.

4/ Write a Runbook and Keying Instructions

Though the runbook is really a series of procedures for the computer operations personnel, it is usually written by the systems analyst instead of a technical writer because of its highly technical content. The same reasoning applies to the key-processing instructions, usually duplicated and included in the runbook. Normally, the runbook consists of a table of contents, job step flowchart, and so forth. For a sample runbook see Figure 4-5a through Figure 4-5l. The same procedure (with appropriately different version) is repeated for every job in the system.

5/ Write Procedures

(See Part 2, Procedures, for specifics on this subject.)

6/ Arrange and Monitor Parallel Processing Operation, or Pilot Operation

In the parallel processing, the new system receives the same data as the old system. This allows the checking of the new system for accuracy at the same time that the old system continues to provide the daily production, and is also a backup. If there is no old system to backup the new system, arrange a pilot operation. This is sort of a "trial run"; only a small percentage of the input being processed at a time.

B/ USER TRAINING/USER'S MANUAL

The purpose of the user training/user's manual task is to define and document every detail of the activities needed to train

SAMPLE OPERATIONS RUNBOOK

BANK PENSION SYSTEM

CROCKER NATIONAL BANK
DATA PROCESSING AND SYSTEMS ADMINISTRATION

TITLE: CONTENTS		Page PROJECT NO.: BPS712
PROJECT: BANK PENSION SYSTEM	PREPARED BY: AN ANALYST	DATE PREPARED: 4/20/74
DEPARTMENT: SYSTEMS & PLANNING	DEPARTMENT SUPERVISOR: A SUPERVISOR	APPROVED BY:

Fig. 4-5a Contents (Courtesy of Crocker Bank)

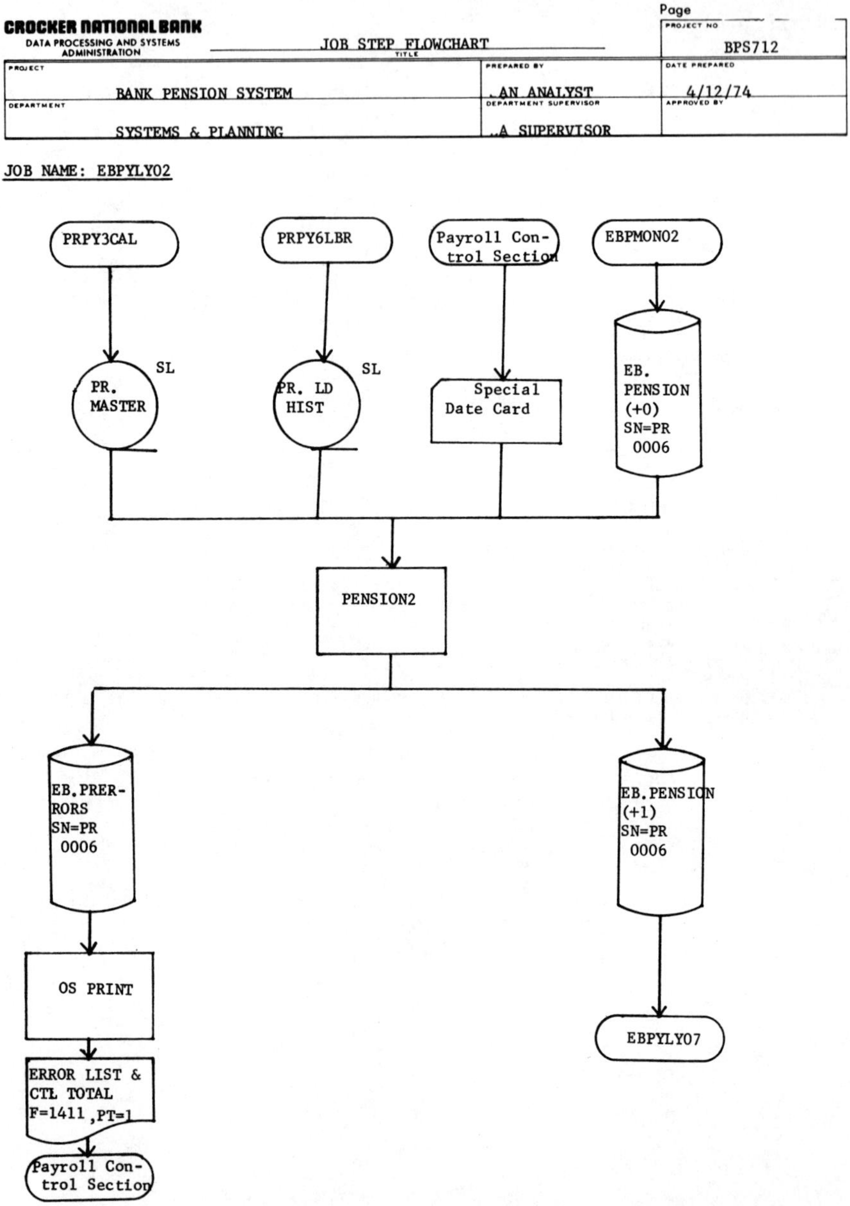

Fig. 4-5b Job Step Flowchart (Courtesy of Crocker Bank)

CROCKER NATIONAL BANK
DATA PROCESSING AND SYSTEMS ADMINISTRATION

TITLE		Page / PROJECT NO.
JOB NARRATIVE DESCRIPTION		BPS712
PROJECT: BANK PENSION SYSTEM	PREPARED BY: AN ANALYST	DATE PREPARED: 4/23/7
DEPARTMENT: SYSTEMS & PLANNING	DEPARTMENT SUPERVISOR: A SUPERVISOR	APPROVED BY:

JOB NAME: EBPYLY02

JOB 4

Step 1. Pension2 updates data from the Date Card (from Payroll Control Section), Payroll Master File, Payroll Labor Distribution History File, and EB.PENSION disk. Step 2. Pension2 generates EB.PRERRORS disk, which is input to OS PRINT utility program, which in turn generates the Error Listing & Control Totals. This printout is routed to Payroll Control Section.

Pension2 also generates EB.PENSION disk which is then input to Job EBPYLY07.

Fig. 4-5c Job Narrative Description (Courtesy of Crocker Bank)

CROCKER NATIONAL BANK
DATA PROCESSING AND SYSTEMS ADMINISTRATION

TITLE		PAGE / PROJECT NO
JCL DECK LISTING		BPS712
PROJECT	PREPARED BY	DATE PREPARED
BANK PENSION SYSTEM	AN ANALYST	4/13/74
DEPARTMENT	DEPARTMENT SUPERVISOR	APPROVED BY
SYSTEMS & PLANNING	A SUPERVISOR	

```
//EBPYLY02 JOB '4532/0000/C00',MASTER,MSGCLASS=1,CLASS=E,PRTY=8,
// TYPRUN=HOLD,REGION=210K
//JOBLIB DD DISP=SHR,DSN=PRDN.SUBTLOAD   **EBPYLY02**
//        DD DISP=SHR,DSN=PRDN.PROGLOAD   **EBPYLY02**
//STP1 EXEC EBPYLY02
//STEP2.CARD DD *
```

```
//STEP1      EXEC PGM=IEHPROGM
//DD1        DD   DISP=SHR,UNIT=SYSDA,VOL=SER=PR0006
//SYSOUT   DD SYSOUT=1
//SYSPRINT DD SYSOUT=1
//SYSIN DD DSNAME=PRDN.CTL(CTLEB005),DISP=SHR
   SCRATCH DSNAME=EB.PRERRORS,VOL=2314=PR0006
   UNCATLG DSNAME=EB.PRERRORS
//STEP2  EXEC  PGM=PENSION2
//SYSABEND  DD SYSOUT=1
//SORTMSG DD SYSOUT=1
//SYSOUT    DD   SYSOUT=1
//SYSPRINT DD SYSOUT=1
//TRANSIN   DD   DUMMY,DCB=BLKSIZE=3200
//PAYIN     DD   DSNAME=PR.MASTER(+0),DISP=OLD
//LABORIN   DD   DSNAME=PR.LDHIST(+0),DISP=OLD
//PENIN     DD   DSNAME=EB.PENSION(+0),DISP=OLD
//PENOUT    DD   DSNAME=EB.PENSION(+1),DISP=(NEW,CATLG,DELETE),
//       DCB=(RECFM=FB,BLKSIZE=3300,LRECL=300),SPACE=(CYL,(25,2)),
//         UNIT=SYSDA,VOL=SER=PR0006
//PRERRORS DD   DSNAME=EB.PRERRORS,DISP=(NEW,CATLG,DELETE),UNIT=SYSDA,
//         DCB=(RECFM=FBA,BLKSIZE=3458,LRECL=133),
//         SPACE=(CYL,(2,2)),VOL=SER=PR0006
//ACTUARY   DD   DUMMY,DCB=BLKSIZE=3458
//DOCUMENT DD   DUMMY,DCB=BLKSIZE=3458
//PRDELETE CD   DUMMY,DCB=BLKSIZE=3458
//ACTOUT    DD   DUMMY,DCB=BLKSIZE=3200
//DISCDOC   DD   DUMMY,DCB=BLKSIZE=3458
//DISCDEL   DD   DUMMY,DCB=BLKSIZE=3458
//SORTLIB   DD    DSN=SYS1.SORTLIB,DISP=SHR
//SORTWK01 DD    UNIT=SYSDA,SPACE=(CYL,(20),,CONTIG)
//SORTWK02 DD    UNIT=(SYSDA,SEP=(SORTWK01)),
// SPACE=(CYL,(20),,CONTIG)
//SORTWK03 DD    UNIT=(SYSDA,SEP=(SORTWK01,SORTWK02)),
// SPACE=(CYL,(20),,CONTIG)
//SORTWK04 DD    UNIT=(SYSDA,SEP=(SORTWK01,SORTWK02,SORTWK03)),
// SPACE=(CYL,(20),,CONTIG)
//SORTWK05 DD    UNIT=(SYSDA,SEP=(SORTWK01,SORTWK02,SORTWK03,SORTWK04)),
// SPACE=(CYL,(20),,CONTIG)
//SORTWK06 DD    UNIT=(SYSDA,SEP=(SORTWK01,SORTWK02,SORTWK03,SORTWK04,
// SORTWK05)),SPACE=(CYL,(20),,CONTIG)
//SYSIN DD DDNAME=CARD
```

```
***DSPRINT***EB.PRERRORS***DISK***PR0006***
```

Fig. 4-5d JCL Deck Listing (Courtesy of Crocker Bank)

CROCKER NATIONAL BANK DATA PROCESSING AND SYSTEMS ADMINISTRATION	TITLE: JOB SETUP PROCEDURES	Page PROJECT NO: BPS712
PROJECT: BANK PENSION SYSTEM	PREPARED BY: AN ANALYST	DATE PREPARED: 1/15/74
DEPARTMENT: SYSTEMS & PLANNING	DEPARTMENT SUPERVISOR: A SUPERVISOR	APPROVED BY:

PROGRAM PENSION2

OPERATION SETUP PROCEDURE FOR THE MONTHLY RUNS

1. Receive SPECIAL DATE Card from Staff Payroll Control Section.

2. Get New PAYROLL MASTER FILE from Staff Payroll System: TRANSACTION FILE from Program PENSION1; BANK PENSION MASTER FILE from DPC Tape Library.

3. Execute PENSION2, which UPDATES data from the Special Date Card and the above magnetic tape files, and generates:

   a. ACTUARY CHANGES Cards
   b. ACTUARY CHANGES File
   c. ERRORS & CONTROL TOTALS File
   d. TURNAROUND DOCUMENTS File
   e, Updated BANK PENSION MASTER FILE.

   The Updated BANK PENSION MASTER FILE replaces the old BANK PENSION MASTER FILE.

4. Input the 3 magnetic tapes: ACTUARY CHANGES File, ERRORS & CONTROL TOTALS File, and TURNAROUND DOCUMENTS File to OS PRINT program.

5. Execute OS PRINT utility program. OS PRINT generates the following monthly reports:

   a. LIST OF ACTUARY PUNCHED CARDS
   b. ERROR LISTING & CONTROL LOTALS
   c. BANK PENSION EMPLOYEE DATA RECORD (TURNAROUND DOCUMENTS)

Fig. 4-5e Job Setup Procedures (Courtesy of Crocker Bank)

CROCKER NATIONAL BANK
DATA PROCESSING AND SYSTEMS ADMINISTRATION

SPECIAL INSTRUCTIONS

TITLE

PROJECT	PREPARED BY	Page PROJECT NO
		BPS712
BANK PENSION SYSTEM	AN ANALYST	DATE PREPARED 4/20/74
DEPARTMENT: SYSTEMS & PLANNING	DEPARTMENT SUPERVISOR: A SUPERVISOR	APPROVED BY

JOB NAME: EBPYLY02

CONSOLE MESSAGE	PROCEDURE
1. MANDATORY ID. OMISSION WILL CAUSE JOB TO ABORT. ID= PENSION2 DATE CARD	Omission of ID. Action: Submit program to Staff Payroll Control Section for correction. Subsequently re-run job.
2. PENSION2 HAS INCORRECT CARD	Incorrect Date Card. Action: Stop run. Insert Standard Date Card and re-run job.
3. PENSION2 NEED MORE CORE FOR COMPANY002. CHANGE PROGRAM HOLDING AREAS	Need more core for Payroll Master File. Action: Stop run. Return job to Staff Payroll Control Section. Notify Programmer.
4. PENSION2 NEED MORE CORE FOR COMPANY 002. CHANGE PROGRAM HOLDING AREAS.	Need more core for Labor History File. Action: Stop run. Return job to Staff Payroll Control Section. Notify programmer.

Fig. 4-5f Special Instructions (Courtesy of Crocker Bank)

CROCKER NATIONAL BANK
DATA PROCESSING AND SYSTEMS ADMINISTRATION

TITLE: LIBRARY TAPE SPECS

Page	
PROJECT NO	BPS712

PROJECT	PREPARED BY	DATE PREPARED
BANK PENSION SYSTEM	AN ANALYST	4/12/74
DEPARTMENT	**DEPARTMENT SUPERVISOR**	**APPROVED BY**
SYSTEMS & PLANNING	A SUPERVISOR	

JOB NAME EBPYLY02

DATA FILE DESCRIPTION

DSNAME/FILE NAME E.B.ACTOUT DDNAME ACTOUT

RETENTION:PERIOD 2 CYCLE

LIBRARY OR TAPE NAME EBACTOUT

REMOTE STORAGE CYCLE Send 1st cycle tape to Remote Storage when 2nd cycle tape is created for 30 days

DSNAME/FILE NAME ______ DDNAME ______

RETENTION:PERIOD ______

LIBRARY OR TAPE NAME ______

REMOTE STORAGE CYCLE ______

DSNAME/FILE NAME ______ DDNAME ______

RETENTION:PERIOD ______

LIBRARY OR TAPE NAME ______

REMOTE STORAGE CYCLE ______

DSNAME/FILE NAME ______ DDNAME ______

RETENTION:PERIOD ______

LIBRARY OR TAPE NAME ______

REMOTE STORAGE CYCLE ______

Fig. 4-5g Library Tape Specs (Courtesy of Crocker Bank)

CROCKER NATIONAL BANK
DATA PROCESSING AND SYSTEMS ADMINISTRATION

TITLE: PRINTER & CARRIAGE TAPE SPECS		PAGE / PROJECT NO: BPS712
PROJECT: BANK PENSION SYSTEM	PREPARED BY: AN ANALYST	DATE PREPARED: 4/12/74
DEPARTMENT: SYSTEMS & PLANNING	DEPARTMENT SUPERVISOR: A SUPERVISOR	APPROVED BY:

JOB NAME EBPYLY02

FORM NAME BANK PENSION EMPLOYEE DATA RECORD FORM NO. 35-676 PLY 3 QUANTITY 500

CARRIAGE TAPE

IDENTIFICATION PENSION2

STANDARD ______ SPECIAL X

CHANNEL LINES PUNCHED

1 7&58

2

3

4

5

6

7

8

9

10

11

12

CUT AND GLUE ON LINE 102

FORMS ALIGNMENT HORIZONTAL

FORMS ALIGNMENT VERTICAL

DISTRIBUTION

☐ UTILITY SECTION

☐ CONTROL DESK

☐ OTHER:

SPECIAL INSTRUCTIONS

☒ OSPRINT

DSNAME EB.DOCUMENT

☐ SYSOUT

EXEC PGM=

DDNAME

Fig. 4-5h Printer and Carriage Tape Specs (Courtesy of Crocker Bank)

JSF

HEADING INFORMATION	WORK STATION	JOB SCHEDULING FORM (JSF)		INFORMATION TYPE			MM	DD	YY	PREPARED BY	PHONE
		PROCESSING CENTRAL PROC	CCN 437	5		WORK ORDER					
		USER BANK STAFF PAYROLL	CCN 542	4		RUN REQUEST					
		APPLICATION BANK PENSION	AN 540	3		DAILY SCHEDULE					
		RUN-NAME EBPYLY02		2	X	PROCESS WORKSHEET	03	15	73	A. SiM	2718
				1		LEAD-TIME WORKSHEET					

SYSTEM ENVIRONMENT	SYSTEM-CODE	OS	SYSTEM-MODE	MVT	SYSTEM-USE	PT
RUN STATUS	DAY EARLY LATE	PRIORITY NORMAL	CURRENT	LATE CAUSE	STATUS CODE	DUE DATE
RUN FREQUENCY	NARRATIVE					
	NFREQ					

JOB IDENTIFICATION		RESOURCES							PRE-REQ/POST-REQ		INPUT / JOB / OUTPUT					ELAPSED HOURS		HOURS TO DUE OUT		SCHEDULED-TIME-DUE				REQUESTED TIME DUE IN/OUT		ACTUAL TIME				OC CODE	ACTUAL ELAPSED HOURS
																				IN		OUT				IN		OUT			
JOB NAME	JOB TITLE	WTR	PCH	RDR	7T TP	9T TP	DK	K	JOB NAME	WORK STATION	I/O	S/D	M	DATA FILE NAME	ADDITIONAL INFORMATION	DELAY	PROCESS	PROCESS	TOTAL	HOUR	DAY	HOUR	DAY	HOUR	DAY	HOUR	DAY	HOUR	DAY		
											I	U	C	PENCARD	DATE CARD																
											I	L	T	PR. MASTER	SN=																
											I	L	T	PR. LDHIST	SN=																
											I	L	D	EB. PENSION (+0)	SN= PR0006																
EBPYLY02	PENSION2	1	1	1		2	2	180	EBPMONO2		J						.2														
											O	L	D	EB. PRERRORS	SN= PR0006																
											O	L	D	EB. PENSION	SN= PR0006																
	PRINT ERROR										I	J	D	EB. PRERRORS	SN= PRC006																
OSPRINT	LIST & CONTROL								EBPYLY02		J																				
	TOTAL										O	U	W	ERROR LISTING	F=1411, PT=1																

RUN – PROCESS FLOW

This JSF is used for the final December run of each year.
Date Card Code = '12'

SHIFT WORKLOAD HOURS			
SHIFT (1)	SHIFT (2)	SHIFT (3)	RUN-TOTAL

JOB NAME	OPERATIONAL DATA
	As-of-Date

Fig. 4-5i Job Scheduling Form (Courtesy of Crocker Bank)

AN ALTERNATE AND SIMPLIFIED JOB SCHEDULING FORM

JOB SCHEDULING FORM (JSF)

USER STOCK TRANSFER CCN 830

APPLICATION CORPORATE TRUST AN

	MM	DD	YY	Prepared by	Phone
RUN REQUEST	08	08	73	J. Broussard	2903

AS OF DATE 08-08-73

CYCLE DATE NONE

RUN – Process Flow.

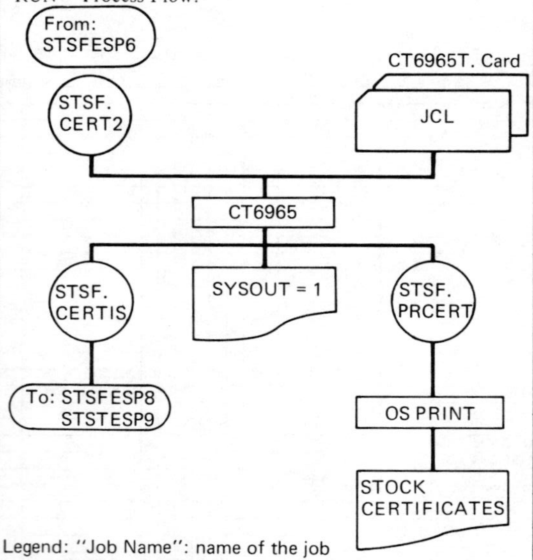

Legend: "Job Name": name of the job to be processed
"K": computer core required to process the job
"T/D/W/C": T = tape, D = Disk, W = writer, C = cards
"Data File Name": actual data set names used in the JCLs.

Job Name	K	I/O	W T/D C	Data File Name	Tape/Disk	Forms	Time Start	Time Stop
STSFESP7	150							
		I	T	STSF.CERT2	TO 5555			
		I	C	CT6965T Call Cards				
		O	T	STSF.CERTIS				
		O	T	STSF.PRCERT				
		O	W	SYSOUT = 1		1412,1pt.		
OSPRINT	30							
		I	T	STSF.PRCERT	TO 4444			
		O	W	STOCK CERTIFICATE		# 1560		

SPECIAL INSTRUCTIONS

Fig. 4-5j Alternate and Simplified Job Scheduling Form (Courtesy of Crocker Bank)

NOTE: DO NOT USE THIS FORM WHEN ORDERING NEW OR REVISED CARD FORM COMPOSITION.

IBM

INTERNATIONAL BUSINESS MACHINES CORPORATION

INFORMATION RECORDS DIVISION
MULTIPLE-CARD LAYOUT FORM

Form X24-6599-1
Printed in U.S.A.

Company: CROCKER NATIONAL BANK

Application: BANK PENSION SYSTEM by AN ANALYST Date 4/10/74 Job No. PENSION2 Sheet No. ____

SPECIAL DATE CARD

CD	As-Of Date	FICA Maxim	Card Identification
01	MMDDYY		PENSION2 DATE CARD
1–2	3–8	9–62	63–80

CD	As-Of Date	FICA Maxim	Card Identification
00	MMDDYY		PENSION2 DATE CARD
1–2	3–8	9–62	63–80

CD	As-Of Date	FICA Maxim	Card Identification
12	MMDDYY		PENSION2 DATE CARD
1–2	3–8	9–62	63–80

Fig. 4-5k Multiple-Card Layout Form (Courtesy of Crocker Bank)

IBM INTERNATIONAL BUSINESS MACHINES CORPORATION

GX24-6299-4 UM/025 *
Printed in U.S.A.

CARD PUNCHING OR VERIFYING INSTRUCTIONS

JOB NAME	JOB NO.	CONTROL PANEL NO.	OPERATION NAME	OP. CODE	MACH. TYPE
BANK PENSION SYSTEM					

FREQUENCY
☐ DAILY ☐ WEEKLY ☐ BI-WEEKLY ☐ SEMI-MONTHLY ☒ MONTHLY ☐ QUARTERLY ☐ ANNUAL ☐ OTHER

DUE IN		DUE OUT		ESTIMATED VOLUME	EST. TIME	
TIME	DATE	TIME	DATE		HOURS	TENTHS

PROGRAM CARD NO. CARD ELECTRO (FORM) NO.

SWITCH SETTINGS - ON
☐ PROGRAM UNIT ☐ PRINT
1 2 ☐☐ PROGRAM SELECT ☐ LEFT ZERO PRINT
☐ AUTO FEED CK PCH ☐☐ SELF CHECK NO.
☐ AUTO SKIP-AUTO DUPL (VER) INS ST ☐☐ CARD INSERT

SPECIAL FEATURES USED
☐ CARD CORRECTION ☐ AUXILIARY DUPLICATE (VERIFY)
☐ CARD INSERTION ☐ SELF CHECK NO.
☒ AUTO LEFT ZERO ☐ SELF CHECK NO. GENERATOR
☐ ALTERNATE PROGRAM ☐ VARIABLE LENGTH FEED
☐ HI SPEED SKIP ☐ INTERSPERSED GANG PUNCH

SOURCE DOCUMENTS USED: BANK PENSION EMPLOYEE DATA RECORD

DISPOSITION OF CARDS:

RECEIVED FROM: EMPLOYEE BENEFITS/PERSONNEL

DOCUMENTS

CARD FIELD	FUNCTION*	COLUMNS FROM	THRU	TOTAL	REMARKS
1. EMPLOYEE NUMBER	PV	1	6	6	
2. SOCIAL SECURITY NUMBER	PV	7	16	10	C.C. 16 may be a blank
3. LAST NAME	PV	17	38	22	Alpha only
4. FIRST NAME INITIAL	PV	39	39	1	Alpha only
5. MIDDLE NAME INITIAL	PV	40	40	1	Alpha or blank only
6. EMPLOYMENT DATE	PV	41	48	8	MM DD YYYY
7. BRANCH NUMBER	PV	49	51	3	
8. DEPARTMENT NUMBER	PV	52	54	3	
9. PAY FREQUENCY CODE	PV	55	55	1	Always "2"
10. PAY RATE (MONTHLY)	PV	56	61	6	Dollar amount only
11. BIRTH DATE	PV	62	69	8	MM DD YYYY
12. SEX	PV	70	70	1	"M" or "F"
13. NEW EMPLOYEE NUMBER	PV	71	76	6	
14. TRANSACTION CODE	PV	77	77	1	"N", "C" or "D"
15. CARD CODE	PV	79	80	2	"20" Always
TOTAL KEY STROKES PER CARD—					

FUNCTION*	SYMBOL
DUPLICATE	D
PUNCH	P
SKIP	S
X-SKIP	XS
VERIFY	V
SELF CK. NO.	CK
LEFT ZERO	LZ

Date: 4/8/74 * No. of forms per pad may vary slightly Section____ Page ____

Fig. 4-51 Card Punching and Verifying Instructions (Courtesy of Crocker Bank)

and orient all level of users to operate the new or revised system. To ensure that all aspects of the system will operate properly it is essential that you monitor this task.

In addition to the procedures (see Part 2), written by the technical writer under the supervision of the systems man, it is also the responsibility of the technical writer to have a complete user's manual ready at the completion of the installation phase to be used as a reference guide by the user. Providing a user's manual is especially important when the system is going to be handled by personnel other than the data processing group. Normally, the user's manual consists of the following sections:

1/ General System Narrative

This section describes briefly and concisely the system and the programs that accomplish the system's objectives. The narrative is illustrated by one or more system flowchart(s).

2/ Input Description

This section explains the type and form of input the user must provide the system, including transaction codes, parameters, program options, and so forth, to get the desired output.

3/ Output Description

This section describes the contents and formats of reports, or any other type of output the system may have.

4/ File Description

This section describes the type and contents of the system files, and the methods by which the master files are updated and maintained.

5/ Operations Instructions

This section is used for reference only, since the detailed procedures for computer operators are in the runbook, and instructions for data processing clerks are in their own procedures book.

A sample application of the user's manual follows.

C/ INSTALLATION

The purpose of the installation task is to define, complete and document the orderly installation of the new or revised system. Throughout this task the systems man carefully monitors the installation, resolving any problems that have not surfaced during system testing and conversion/installation planning. To ensure that all activities required within this task have been completed, prepare the installation checklist (Figure 4-6). As a final step of the installation phase, fill out the implementation results log (Figure 4-7). This form provides a continuous chronologic record of system performance during this phase.

CROCKER NATIONAL BANK DATA PROCESSING AND SYSTEMS ADMINISTRATION	TITLE: INSTALLATION CHECKLIST		Page PROJECT NO. TDS1234
PROJECT: TIME DEPOSIT SYSTEM		PREPARED BY: AN ANALYST	DATE PREPARED: 4-25-74
DEPARTMENT: BANK OPERATIONS		DEPARTMENT SUPERVISOR: A SUPERVISOR	APPROVED BY:

Tasks	Date Completed
Setup remote storage for	
generation of files	4-10-74
Test production JCL	4-12-74
Check supply of preprinted forms on hand	4-12-74
Prepare information for Documentation/	
Control Library	
1. Program Distribution Advices form	4-15-74
2. Post List, including service statement	
showing System Information Date & Compile	4-22-74
3. Instructions as listed in the computer Runbook	4-23-74
4. Complete Maintenance Log	4-24-74
5. List of all modules and data definitions	4-25-74
6. Notify Library of implementation date	4-15-74

Fig. 4-6 Installation Checklist (Courtesy of Crocker Bank)

CROCKER NATIONAL BANK
DATA PROCESSING AND SYSTEMS ADMINISTRATION

RESULTS LOG

☐ SYSTEM TEST ☒ IMPLEMENTATION

Page
PROJECT NO. TDS1234

PROJECT	PREPARED BY	DATE PREPARED
TIME DEPOSIT SYSTEM	AN ANALYST	4-10-74
DEPARTMENT	DEPARTMENT SUPERVISOR	APPROVED BY
BANK OPERATIONS	A SUPERVISOR	

DATE OF TEST	TIME OF TEST	TEST	OPERATOR RESPONSIBLE	ANALYST RESPONSIBLE	EA ✓	RESULTS
4/8	4 a.m.	POSTING	#6	Smith		Branch Priority Table not created for use on this run.
4/9		REPORT ORGANIZER	#2	Smith		Inactive Totals on BTR* incorrect for Branch 158
4/10		SCHEDULER	#6	Smith		Scheduler data set created with card utility to get through the run
* Branch Transaction Register						

Fig. 4-7 Results Log–Implementation (Courtesy of Crocker Bank)

SAMPLE APPLICATION

DOCUMENT CONTROL SYSTEM

USER'S MANUAL

April 1974

BOULDER ENGINEERING, INC.

CONTENTS

Section I

GENERAL SYSTEM NARRATIVE

The Document Control System was designed to control the Company's Engineering Change Orders (ECO) and all documents. The system is implemented on the Burroughs B3500 computer, under Disk File Master Control Program (DF MCP). It runs daily to maintain an up-to-date ECO and document files, and prints 14 daily, weekly and monthly reports, and exception reports on request.

There are five programs in the Document Control System, and each performs a specific function. The first program collects and sorts all ECO and document input data (Figure B-1a). The second program updates the ECO Master File (Figure B-1b). The third program sorts document input data passed on to it by the second program (Figure B-1c). The fourth program updates the document Master File with document maintenance and ECO information (Figure B-1d). The fifth and last program sorts and prints reports generated in the second and fourth program (Figure B-1e).

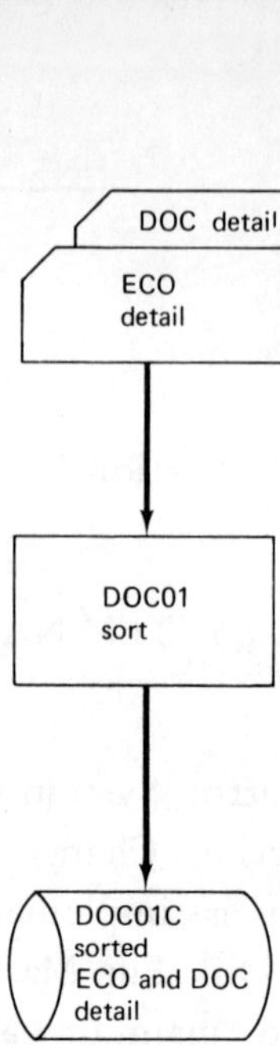

Fig. B-1a

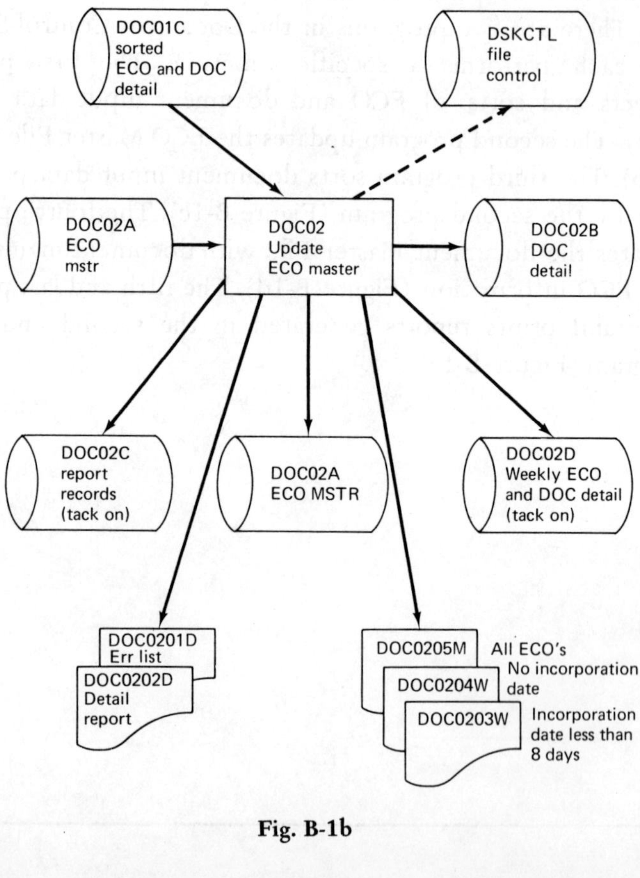

Fig. B-1b

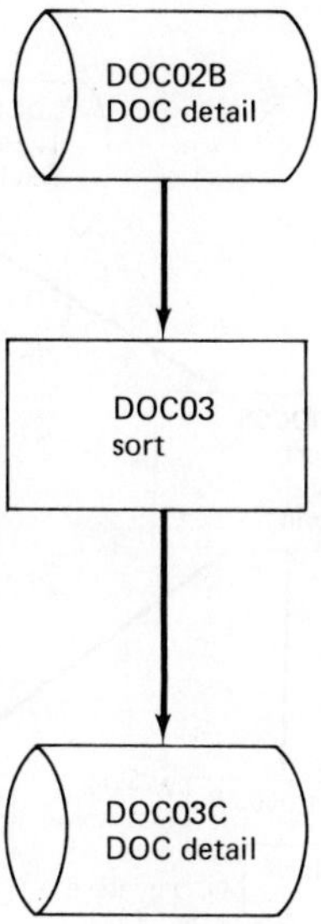

Fig. B-1c

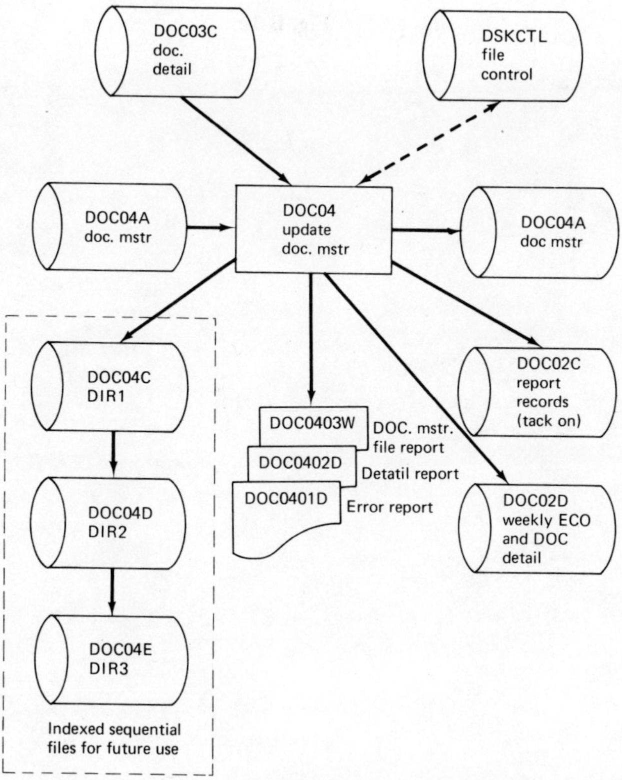

Fig. B-1d

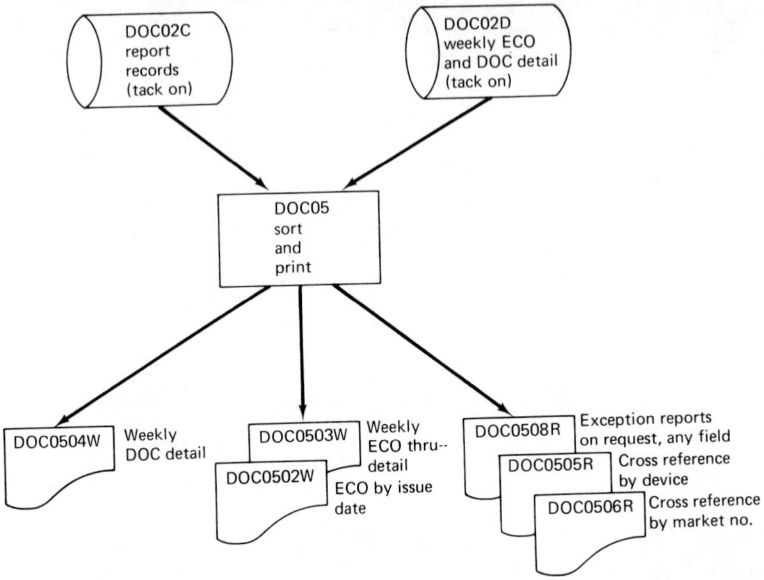
DOC02C
report
records
(tack on)
DOC02D
weekly ECO
and DOC detail
(tack on)
DOC05
sort
and
print
DOC0504W
Weekly
DOC detail
DOC0503W
Weekly
ECO thru--
detail
DOC0502W
ECO by issue
date
DOC0508R
Exception reports
on request, any field
DOC0505R
Cross reference
by device
DOC0506R
Cross reference
by market no.

Fig. B-1e

Section II

INPUT DESCRIPTION

There are 14 transactions to maintain the Document Control System. All transactions are manually input on the appropriate forms by the Document Control Department. All transactions have a minimum of transaction code and document control number for control purposes. The document control numbers consist of basic device, serial, function code and sequence numbers.

A/ EXAMPLES OF INPUT

To ensure valid input, the following examples are the correct ways for filling out the various input forms for all transactions.

- Example 1–Set up an issued ECO (Engineering Change Order)
- Example 2–Set up a released ECO
- Example 3–Delete a complete ECO
- Example 4–Delete ECO item or items
- Example 5–Change ECO item or items
- Example 6–Release all ECO document revisions and treat them as changes to the document file
- Example 7–Set up basic specifications on document file
- Example 8–Set up customer specifications
- Example 9–Set up drawings, manufacturing transfer, manufacturing acceptance, probe card, test equipment and programming aids, and customer test equipment

- Example 10–Delete a document record from the document file
- Example 11–Change a document record
- Example 12–Add document release data and/or status
- Example 13–Change document release date and/or status
- Example 14–Request weekly, monthly and cross reference or exception reports.

B/ PARAMETER

There is one parameter card in this system. The parameter card is used to request weekly, monthly and cross reference or exception reports. (Example 14)

Transaction Codes

200/201= DELETE
202/203= ISSUE
204/205= RELEASE
206/207= INCORPORATE
208/209= CHANGE

ECO INPUT

TRC	ECO NUMBER (DEVICE, SER, FN, SEQ)	CONT	DOC AND REV (SER, FN, SEQ)	REV	ISS. DATE (MO DA YR)	REL. DATE (MO DA YR)	DIST. DATE (MO DA YR)	INC. DATE (MO DA YR)	REASON FOR CHANGE
1–3	4–15	16	17–25		26–31	32–37	38–43	44–49	50–74
202	005680005201		X		021572				CØRR DSGN RULE VIØLATIØN
203			00021						
			0004o						

Fig. B-2a Example 1

ECO INPUT

TRC	ECO NUMBER (DEVICE, SER, FN, SEQ)	CONT	DOC AND REV (SER, FN, SEQ)	REV	ISS. DATE (MO DA YR)	REL. DATE (MO DA YR)	DIST. DATE (MO DA YR)	INC. DATE (MO DA YR)	REASON FOR CHANGE
1–3	4–15	16	17–25		26–31	32–37	38–43	44–49	50–74
204	005680005201		X			021972			CØRR DSGN RULE VIØLATIØN
205			00021	A					
			00040	A					

Fig. B-2b Example 2

ECO INPUT

TRC	ECO NUMBER (DEVICE / SER / FN / SEQ)	CONT	DOC AND REV (SER / FN / SEQ / REV)	ISS. DATE (MO / DA / YR)	REL. DATE (MO / DA / YR)	DIST. DATE (MO / DA / YR)	INC. DATE (MO / DA / YR)	REASON FOR CHANGE
1–3	4–15	16	17–25	26–31	32–37	38–43	44–49	50–74
200	005680005201	D						
201			00021					

Fig. B-2c Example 3

ECO INPUT

TRC	ECO NUMBER (DEVICE / SER / FN / SEQ)	CONT	DOC AND REV (SER / FN / SEQ / REV)	ISS. DATE (MO / DA / YR)	REL. DATE (MO / DA / YR)	DIST. DATE (MO / DA / YR)	INC. DATE (MO / DA / YR)	REASON FOR CHANGE
1–3	4–15	16	17–25	26–31	32–37	38–43	44–49	50–74
200	005680005201							
201			00021					
			00040					

Fig. B-2d Example 4

ECO INPUT

TRC	ECO NUMBER				CONT	DOC AND REV				ISS. DATE			REL. DATE			DIST. DATE			INC. DATE			REASON FOR CHANGE
	DEVICE	SER	FN	SEQ		SER	FN	SEQ	REV	MO	DA	YR	MO	DA	YR	MO	DA	YR	MO	DA	YR	
208	005680005201																					
209						00021			B													

Fig. B-2e Example 5

ECO INPUT

TRC	ECO NUMBER				CONT	DOC AND REV				ISS. DATE			REL. DATE			DIST. DATE			INC. DATE			REASON FOR CHANGE
	DEVICE	SER	FN	SEQ		SER	FN	SEQ	REV	MO	DA	YR	MO	DA	YR	MO	DA	YR	MO	DA	YR	
204	005680005201				R																	
205						00021			A													

Fig. B-2f Example 6

BASIC SPEC, MASK, DRAWINGS
RUBIES, SECTIONS, AND FUNCTIONS 53 THRU 63

210 DELETE
211 SET UP
212 CHANGE

Fig. B-2g Example 7

TRC	DOC CTL NO.				REL DATE			JOB ORDER NO.	DEV LOC	DIE SIZE	NUM PADS	DEVICE COUNT	HORA	DOC SZ	CLASS	PROC	RUBY SIZE	DESIGN RULES	REV	SORF	
	DEVICE	SER	FN	SEQ	MO	DA	YR														
1–3	4–8	9–11	12–13	14–15	16–17	18–19	20–21	22–29	30–32	33–38	39–41	42–46	47	48	49	50–51	52–55	56–67	68	69	70–80
211	00568	000	40		05	04	72	05100025	BEI	095095	026	329	H	B	A	D	200X	ABCDEF			

CUSTOMER SPECIFICATION

210 DELETE
211 SET UP
212 CHANGE

Fig. B-2h Example 8

TRC	DOC CTL NO.				MARKET NUMBER	CCN	WORK STATEMENT	PKG CODE	SUPERSEDES		SUPERSEDED		DEVICE DESCRIPTION	CUSTOMER NAME
	DEVICE	SER	FN	SEQ					DEVICE	SER	DEVICE	SER		
1–3	4–8	9–11	12–13	14–15	16–21	22–24	25–31	32–34	35–39	40–42	43–47	48–50	51–66	67–80
211	00568	001	40		C01320	071	2038	IBX	00318	000			REDØ 0318	BEI R/D

BASIC SPEC, MASK, DRAWINGS
RUBIES, SECTIONS, AND FUNCTIONS 53 THRU 63

210 DELETE
211 SET UP
212 CHANGE

TRC	DOC CTL NO.				REL DATE			JOB ORDER NO.	DEV LOC	DIE SIZE	NUM PADS	DEVICE COUNT	HORA	DOC SZ	CLASS	PROC	RUBY SIZE	DESIGN RULES		SORF	
	DEVICE	SER	FN	SEQ	MO	DA	YR												REV		
1–3	4–8	9–11	12–13	14–15	16–17	18–19	20–21	22–29	30–32	33–38	39–41	42–46	47	48	49	50–51	52–55	56–67	68	69	70–80
211	00568	000	01		05	04	72														
211	00568	000	21		05	04	72														
211	00568	000	32		05	04	72														
211	00568	000	41		05	04	72														
211	00568	000	53		05	04	72														
211	00568	000	54		05	04	72														
211	00568	000	60		05	04	72														
211	00568	000	61		05	04	72													S	
211	00568	001	62		05	04	72													S	
211	00568	001	63		05	04	72														

Fig. B-2i Example 9

BASIC SPEC, MASK, DRAWINGS
RUBIES, SECTIONS, AND FUNCT'ONS 53 THRU 63

210 DELETE
211 SET UP
212 CHANGE

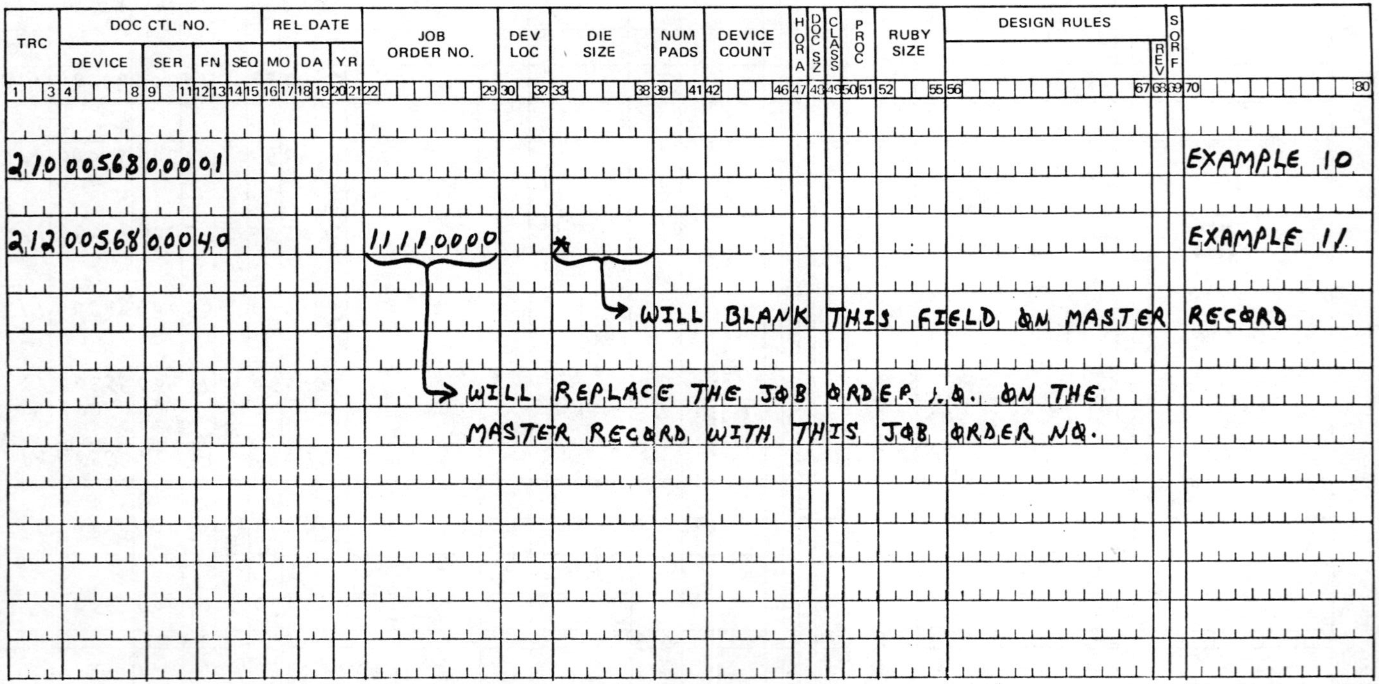

Fig. B-2j Examples 10 and 11

DOCUMENT RELEASE DATE AND/OR STATUS

213 ADD REL DATE AND/OR STATUS
214 CHANGE REL DATE AND/OR STATUS

TRC	DOC CTL NO DEVICE	SER	FN	SEQ	REL DATE MO	DA	YR	STAT CD	
1–3	4–8	9–11	12–13	14–15	16		21	22	23–80
213	00568	000	40		05	10	72		
213	00568	000	40					I	WILL CODE COMPLETE DEVICE ON DOCUMENT MASTER FILE AS INACTIVE AND DELETE ALL DOCUMENT RECORDS FOR THIS DEVICE EXCEPT BASIC AND CUSTOMER SPECIFICATIONS.

Fig. B-2k Example 12

DOCUMENT RELEASE DATE AND/OR STATUS

213 ADD REL DATE AND/OR STATUS
214 CHANGE REL DATE AND/OR STATUS

TRC	DOC CTL NO: DEVICE	SER	FN	SEQ	REL DATE: MO	DA	YR	STAT CD	
1–3	4–8	9–11	12–13	14–15	16		–21	22	23–80
214	00568	000	40		05	15	72		
214	00568	000	40					I	} THE SAME AS EXAMPLE 13 OR TRANSACTION 213

PLACE X IN FIELD OR FIELDS FOR WHICH REPORTS ARE DESIRED

Fig. B-21 Example 13

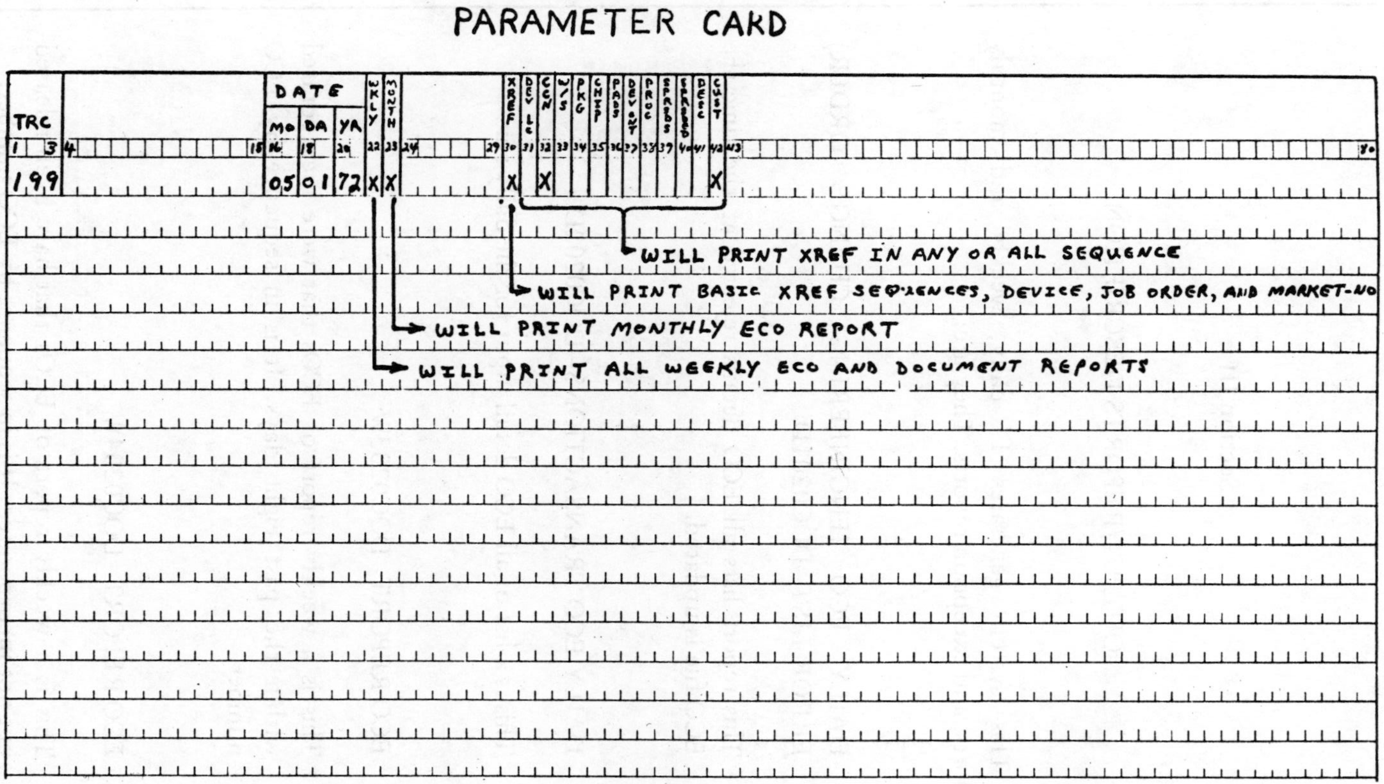

Fig. B-2m Example 14

Section III

OUTPUT/REPORTS DESCRIPTION

This system generates 14 daily, weekly and monthly reports, and exception reports. They are:

1/ DAILY ECO (ENGINEERING CHANGE ORDER) ERROR LIST–DOC0201D

This report lists all ECO details rejected at the time the ECO file is updated.

2/ DAILY ECO TRANSACTIONS–DOC0202D

This is a list of all ECO detail input for the daily run.

3/ ECO REPORT–DOC0203W

This is a weekly report of ECOs that were incorporated within the past eight days. It is in sequence by ECO number.

4/ ECO REPORT–DOC0204W

This is a weekly report of ECOs that have been released but not incorporated. It is in sequence by ECO number.

5/ ECO MASTER FILE REPORT–DOC0205M

This is a monthly report of all ECOs that are currently on the ECO Master File. It is in sequence by ECO number.

6/ ECO REPORT–DOC0502W

This is a weekly report of ECOs issued but not released. It is in sequence by issue date.

7/ WEEKLY ECO TRANSACTIONS–DOC0503W

This report is a listing of the daily ECOs submitted for the week. It is in sequence by ECO number.

8/ DOCUMENT DETAIL ERROR LIST–DOC0401D

This daily report lists all document details rejected at the time the document Master File was updated.

9/ DAILY DOCUMENT DETAIL–DOC0402D

This report lists all document details processed.

10/ DOCUMENT MASTER FILE–DOC0403W

This is a weekly report of all documents that are currently on the Document Master File. It is in sequence by document control number.

11/ WEEKLY DOCUMENT DETAIL–DOC0504W

This report lists all daily documents submitted for the week. It is in sequence by document control number.

12/ CROSS REFERENCE BY DOCUMENT CONTROL NUMBER–DOC0505R

This report contains device and customer information taken from all documents related to a device. It is in sequence by document control number.

13/ CROSS REFERENCE BY MARKETING NUMBER–DOC0506R

This report contains the same information as DOC0505R, but is in sequence by marketing number.

14/ EXCEPTION REPORT

This report is a cross reference, and is in the same format as the weekly cross reference. The exception cross reference can be in sequence by any one of the fields that appear on the report.

An example of a weekly report–ECO REPORT–DOC0203W–follows.

DOC0203W

E C O   R E P O R T

(INCORP. DATE LESS THAN 8 DAYS)

REPORT DATE 1/31/72 PAGE NO. 1

ECO NUMBER	ISSUED DATE	RELEASED DATE	DISTR DATE	INCORP DATE	REASON FOR CHANGE	DOCUMENTS REVISED
00584-000-52-00		1/26/72	1/31/72	1/25/72	SET UP CORRECT REVISIONS	40-A 42-A 001-40-E 002-40-A 003-40-E 004-40-A

Fig. B-3 ECO Report

Section IV

FILE DESCRIPTION

The Document Control System consists of two master files: ECO Master File and Document Master File. These files are updated daily by manual input from the Document Control Department.

These files are:

A/ ECO MASTER FILE–DOC02A

This master file is updated with add, delete or change transactions 200 through 209. These transactions are submitted on BEI Form 502.

B/ DOCUMENT MASTER FILE–DOC04A

This master file is updated with add, delete or change transactions 210 through 219. Transactions 210 through 214 are submitted on BEI Forms 502-1–BEI 502-5. Transactions 215 through 219 are internal transactions generated from the ECO Master File.

Figures B-4a and B-4b are sample formats of master records used for control of ECO and document records.

RECORD LAYOUT

PROGRAM TITLE	FILE NAME	JOB I.D.	CDP FILE I.D.
DOCUMENT CONTROL	ECO MSTR	DOC	DOC02A

LENGTH	RECORD NO. PER BLOCK	SEQUENCE FIELDS (major to minor)
80	10	DEVICE, ECO NO., SERIAL, FUNCTION, SEQNCE.

FILE SEQUENCE: [X] ASCEND'G [ ] DESCEND'G [ ] RANDOM
LABELS: [X] STANDARD [ ] NON-STANDARD [ ] NONE

INPUT TO	OUTPUT FROM
DOC02	DOC02

PADDING: [ ] NINES [X] BLANK [ ] OTHER (SEE COMMENTS)
MODE: [ ] BINARY [X] BCD
DENSITY: [X] HIGH [ ] LOW
COLLATING: [X] COMM. [ ] SCIENTIFIC

ECO CTL						ISSUE DATE			REL DATE			DIST DATE			INCORP DT.			
DEVICE	ECO	SER	FUNCT	SEQ	REV	MO	DY	YR	MO	DY	YR	MO	DAY	YR	MO	DAY	YR	REASON FOR CHANGE
9(5)	99	999	99	XX	XX	99	99	99	99	99	99	99	99	99	99	99	99	X(25)

6 12 18 24 30 36 42 48 54 60

	FILLER X(15)

66 72 78 84 90 96 102 108 114 120

126 132 138 144 150 156 162 168 174 180

COMMENTS ECO file updated daily. ECO's may be added, changed, or deleted. Major control is by device, ECO-No., serial, function and sequence.

ALL-PURPOSE BOND DIAZO REPRODUCIBLE

PREPARED BY ________ DATE ________

PAGE 1 OF 1

Fig. B-4a Record Layout ECO Master File

RECORD LAYOUT

PROGRAM TITLE	FILE NAME	JOB I.D.	COP FILE I.D.
DOCUMENT CONTROL	DOC MSTR	DOC	DOC04 A

RECORD LENGTH	NO. PER BLOCK	SEQUENCE FIELDS (major to minor)	FILE SEQUENCE	LABELS
300	5	DEVICE, SERIAL, FUNCTION, SEQUENCE	[X] ASCEND'G [ ] DESCEND'G [ ] RANDOM	[X] STANDARD [ ] NON-STANDARD [ ] NONE

INPUT TO	OUTPUT FROM	PADDING	MODE	DENSITY	COLLATING
DOC04	DOC04	[ ] NINES [X] BLANK [ ] OTHER (SEE COMMENTS)	[ ] BINARY [X] BCD	[X] HIGH [ ] LOW	[X] COMM. [ ] SCIENTIFIC

CHECK (✓) IF THIS RECORD IS A CONTINUATION FROM ANOTHER SHEET

DOC. CTL NO.								REL DATE			DIST DATE			
DEVICE	SER	FUNCT	SEQ	ADDRESS	DOC REV	ECO NO.	PEND ECO NO.	MO	DY	YR	MO	DY	YR	REASON FOR CHANGE
9(5)	999	99	XX	9(6)	XX	99	99	99	99	99	99	99	99	X (25)

Positions 1–60

FILLER
X (6)

Positions 61–120

Positions 121–180

CHECK (✓) IF THIS RECORD IS CONTINUED TO ANOTHER SHEET

COMMENTS DOC file updated daily. Documents may be added, changed, or deleted. Major control is by device, serial, function, and sequence.

ALL-PURPOSE BOND DIAZO REPRODUCIBLE

PREPARED BY ________ DATE ________

PAGE 1 OF 1

Fig. B-4b Record Layout Document Master File

Section V

OPERATIONS INSTRUCTIONS

A/ TIME OF INPUT

All input must be in the Keypunch Section *before* 5:00 p.m., daily.

B/ KEYPUNCH INSTRUCTIONS

Input forms are to be punched as submitted. Keypunching and verifying must be completed in time for program to be run on swing shift.

C/ PROCESSING SCHEDULE

Keypunched input are to be inserted behind data card "DOC01A" in program DOC01. Programs have to be run on swing shift so that reports are available the following morning.

D/ PROGRAMS IN THE SYSTEM

- DOC01–Sorts all input data
- DOC02–Updates ECO file
- DOC03–Sorts document input data
- DOC04–Updates document file
- DOC05–Sorts and prints reports

E/ SYSTEM FLOW

- Load DOC program and data files from latest "DOCJRK" tape. Execute DOC01. Program DOC02 will be "zipped" from DOC01. ("Zipping" or chaining of programs from one program to the next is often possible on the Burroughs System.)
- Program DOC02–zips DOC03.
- Program DOC03–zips DOC04.
- Program DOC04 will zip library dump to tape "DOCJRK" if program DOC05 does not run. DOC05 will not run unless weekly or monthly reports are requested. If DOC05 is going to run, program DOC04 zips DOC05 only.
- Program DOC05 zips library dump to tape "DOCJRK."

F/ DISPOSITION OF REPORTS

It is the daily responsibility of Document Control Department to pick up the output/reports.

G/ RETENTION OF OUTPUT FILES

Disk files dumped to tape (DOCJRK) in program DOC04 or DOC05 are to be retained for one week or five runs.

H/ RESTART PROCEDURE

Mount the latest library tape (DOCJRK) and re-execute DOC01. When accept message appears in programs DOC02 and DOC04, give an "OK."

Chapter

5

Postinstallation Phase

This chapter describes the type and amount of information the systems man must provide the auditor to have the new or revised system audited and evaluated. The auditor also establishes control criteria for the accounting data that are external to the computer, such as checking and verifying preparation of input data and distribution of output data; as well as internal to the computer, such as controls designed into the system.

The last part of this chapter gives detail instructions and an example of a technical report. This is the final report the systems man submits to the user/management, reporting the results of the implemented system.

The postinstallation phase consists of two tasks: A/ auditing and evaluating and B/ final report.

A/ AUDITING/EVALUATING

There are two types of auditors: company and outside or government auditors. Both kinds are concerned with the control and accuracy of business information. For the company auditor to audit and evaluate the new or revised system for effectiveness, reliability and timeliness he needs very specific material from the documented system. It is the systems analyst's responsibility to ensure that the auditor is provided a functional audit control manual compiled by the technical writer.

The audit control manual, to be useful to the auditor, must contain the following minimum information duplicated from the appropriate sections of the documented system.

- Input preparation procedures that verify input data accuracy.
- Source documents, their functions and methods of verification after the information on the source documents are converted via keypunch, key-to-disk/tape/cartridge or on-line terminals into computer-acceptable input to the system. In other words, the exact techniques used for input controls on source documents whether these are engineering change orders, stock transfer data, sales orders, personnel change orders or whatever. One technique of input control, for example, is data keyed in twice, so that the second input can be verified against the first input. There are many other methods.
- Program(s) which in the system capture the original input, and program(s) which capture the reentry input.
- Data that are passed on to other systems, if there are other systems.
- Type of control records the system contains. How the data in the system are checked for accuracy and how errors are corrected.
- Edit controls, i.e., is there an error listing programmed into the system, so that any errors that occur during the SORT/EDIT program are not only printed out but analyzed. Such report is a control measure for the validity of input data after they have been sorted and edited.
- File controls that are designed into the system (generated by particular programs) and that can be used as audit checkpoints. One effective file control method is the "proof totals," whereby certain files are checked for accuracy at the end of the day's processing. Thus, for example, in an inventory control system the control records may be "Add," "Delete" and "Change." These controls are then compared between SORT and EDIT programs, and between SORT and

MERGE programs, and the totals must agree. If they do not balance they are printed out on the exception report.

- Output controls that can be achieved by well-designed control reports containing some type of balancing procedures. For example, the control report that lists the daily runs and also gives statistics about the records in the master files *before* and *after* it is updated. If all the figures do not agree the data appear on the exception report. Another output control technique is the "cross-index" report. In this type of report the horizontally listed data must agree with the vertically listed data, otherwise the discrepancy appears on the exception report.
- Error and reentry procedures that are standardized. If the systems analyst is to get the auditor's approval, he has to establish standards for the processing of errors and the reentering of data when he designs the exception reports within the system. Such reports should show not only the invalid transactions of each run but also the reason the items were not processed, and also the error messages.
- General system flowcharts and job step flowcharts that are essential to the auditor to review and check the new or revised system.

B/ FINAL REPORT

The purpose of this task is to write a report to the user/management on the progress and results of the installed system. The report should reflect the careful consideration the systems man gave to the user's requirement. It also should describe the system approach used in solving the problem, some of the problems encountered, the goals achieved and the built-in system flexibility for possible future modification or expansion.

The technical report, if it is to communicate effectively, has to be well organized and clearly written in a direct, uncomplicated style. The format of the technical report is comprised of three parts: 1/ front matter, 2/ body and 3/ back matter.

1/ Front Matter

This part consists of (1) front cover, (2) title page, (3) foreword or preface (optional), (4) abstract or summary (optional) and (5) table of contents. (See sample report.)

2/ Body

This, the main part of the technical report, consists of:

a/ *Introduction*

This includes very brief background information and possibly early history of the project.

b/ *Objectives/Goals*

This should be in accordance with the proposal and the user requirements.

c/ *Discussion*

This consists of describing briefly the problems encountered, file structure (optional), a command or query language, input/output and the result of the implementation of the new or revised system.

d/ *Conclusion*

This summarizes the goals achieved and informs the user/management of the extent of the designed system's flexibility for future changes or expansion.

3/ Back Matter

The last part of the technical report consists of (1) appendixes (optional) and (2) back cover.

The sample technical report follows.

SAMPLE APPLICATION

FINAL REPORT ON THE LIBRARY INFORMATION SYSTEM

April 30, 1974

BOULDER ENGINEERING, INC.

CONTENTS

Final Report

LIBRARY INFORMATION SYSTEM

A/ INTRODUCTION

In response to Boulder Engineering scientific and engineering staff's urgent need for an improved information service through a computer system to search document collections, in-house preprints, engineering/scientific journals, articles, and so forth, the Library Information System ("LIZ" for ease of pronounciation and recall), a time-shared, on-line retrieval system was designed and developed. Prior to the design phase, however, a comprehensive system study was conducted. This undertaking included making a lexicon of data to see how many staff are using the same data and how these multiple users are going to use the data. Consequently, LIZ is not just a new automated system replacing the old and rather cumbersome manual system. It is a less tedious and more efficient tool for both the library and the engineering/scientific staff. It allows bibliographic and literary research to be performed quickly and without drudgery.

Primarily LIZ is an information retrieval system with the capability to let the user search a data base to gather selected information via the CRT (cathode ray tube) display communication terminal. In addition, LIZ outputs requested documents, and permits the user to control the session at the terminal. To handle these basic functions, LIZ has four corresponding modules (groups of programs). They are: SEARCH, COLLECT, RETRIEVE and OUTPUT. These modules in turn are supervised by the

MONITOR, a master control program. It is the MONITOR's responsibility to ensure that the user is switched from one module to another; each module operating as if it were alone in the world. The MONITOR supervises the interfacing of LIZ with the company's other systems.

The nonprogrammer using LIZ is not and has no need to be aware which module he is operating in. All he has to do is learn the above four modules and their legal commands. By keying in any of these modules, the user explicitly transfers to its control.

B/ OBJECTIVES

To ensure that the new system will be responsive to the user's requirements, the following objectives were strictly adhered to in designing LIZ:

- Provide a system that has the capability to create the update mass storage data files.
- Provide a system that has the capability to search (to have direct access to) selected portions of stored data files. Moreover, to retrieve these data via a simple command language to be examined on the screen of the CRT, or to be output as "hard copy" via the printer or the terminal.
- Above all, to provide a reliable, flexible system.

Coupled with these objectives was a deliberate attempt to avoid designing a system that would be too complex or too expensive to be installed and used.

Now that LIZ has been tested and successfully implemented, the librarian-programmer has taken over the responsibility to create and maintain all data collections. The user (engineers, scientists, librarians, library clerks, etc.), by defining such data as the author's name, journal title, etc., may search various sections of the stored data files and thus gather and retrieve data in any combination.

C/ DISCUSSION

1/ Decisions

During the development of LIZ many decisions had to be made. Here are the major ones:

a/ *Go to batch processing or use time-shared system?*

Admittedly, a batch processing system is cheaper. However, the slow feedback characteristics of batch-processed output would not have been much improvement over the present tedious manual system. Consequently, the choice was for an interactive, time-shared system that responds immediately to requests for information.

b/ *Design a general system that can provide Boulder Engineering a variety of services, e.g., bibliographic, document control, personnel, etc? Or design a special purpose system that provides bibliographic information alone?*

Apart from economics, it was found that special purpose systems, dedicated to one type of service, are usually more efficient than general systems. Consequently, it was decided to make LIZ a dedicated system. However, because of its built-in flexibility, once the system is established and its performance is consistently reliable, adding other types of records and output formats for inventory control, personnel files, etc., at some time in the future will involve only relatively small programming effort.

c/ *Should LIZ provide bibliographic information alone, or should the system be expanded to provide full text retrieval from data files of nonbibliographic records?*

The final decision was that at this time LIZ should provide only bibliographic data. However, this does not mean that in the future it cannot be expanded to go beyond engineering/scientific references, and provide full text of some document on the CRT or printout.

2/ File Organization

In searching for the proper file organization—a major task in any information retrieval system—the following factors had to be considered: speed of search, costs of input and update, ease of modifying files, and storage media. After deliberating techniques such as the tree structure, network structure, etc., the simple, yet effective inverted file organization was decided upon. Thus, data are input serially without any imposed order or organization, and are accessed by indexes. All structuring is done in the multiple-level index files. Since "hashing technique" (finding a record address via a given key) and "chaining" (using index or random technique) through a list of "keys" (special attributes to select a record from the file) in a small, specified segment of an index file are of concern only to systems designers and programmers and not to the user, it is sufficient to say that the file organization of LIZ utilizes the company's

IBM System 360, Model 65, OS (Operating System), and minimizes the number of accesses to the two 2311 disk drives, our storage media.

3/ Command/Query Language

The objectives dictated the programming language, COBOL (all programs are written in COBOL, except a few routines that are in Assembly Language), and the command language for LIZ. In other words, an uncomplicated query language is provided to the user, so that he/she can give simple, unambiguous commands to the computer for effective response. These commands are in a language as close as possible to natural English. It consists of names of index files that are to be searched, followed by the value of specifics to be searched, e.g., AUTHOR: Van Duyn, TITLE: Documentation Manual. Logical operators such as "and," "or," "if," "not," etc., are also part of the command language when the user does a more complex search. Even in such cases the user has no need to know where in the hierarchical network he is working; the MONITOR is structured so that it supervises such details also.

4/ Input/Update

In any information system it is crucial how the data are input, verified, corrected and updated before sent into the data files. In the case of LIZ, the Library of Congress is providing the initial bibliographic data on magnetic tapes. However, all the additional data are going to be input, verified, corrected and updated—i.e., maintained—via the CRT terminal.

D/ CONCLUSION

LIZ by no means is an optimal computer system; it was not intended to be one. Rather, LIZ is an attempt to provide a computer system that is adapted to the needs of the nonprogrammer user, instead of vice versa. Aside from getting bibliographic information faster and without the usual tedious manual research, the greatest benefit of LIZ is that multiple users can examine the same data from the same data collection simultaneously, something quite impossible with the manual system. Also, via the CRT display terminals, the user may study information stored on microfiche as well as information stored in the data files.

Nevertheless, it is planned that in the future LIZ will provide not only full text of various documents but also the capability to inform the user (upon request) when new data have been input (add, delete, change) that affect the status of a particular subject.

Whether these plans or others are going to be implemented will depend on many factors, the most important being user response and economics. However, LIZ was designed with a built-in flexibility so that any future modifications or expansions may be accomplished without any redesigning of the system.

Part 2

Procedures

Most companies recognize the need for clear, well-organized and simply illustrated standard procedures. So often, however, these same companies will allow their technical or administrative writers to do "their own thing" insofar as format, style and convention of the company procedures are concerned. Invariably the result is chaos.

Moreover, a surprisingly large percentage of businesses whose success or failure depends on efficient operation, or manufacturing, or service, or processing of paper work, have no written procedures for the various tasks, and rely on "knowledgeable" employees.

Written standard procedures are imperative for uniformity and efficiency; for ease of learning and understanding whatever type of work the new or the transferred employee is told to do. The larger the company, the more the need exists to establish appropriate written standards for procedures. This is true whether the procedures are for preparing data for computer input, manually producing a product or giving vital service to the company. The keyword is: APPROPRIATE. Consequently, this part of the manual presents an array of various types of standard procedures.

Because no two companies—not even in the same field—operate the same way, it is a management decision as to which type/format/style of procedure is most appropriate for the particular facility. Once the decision is made the selected type of standard procedures should be established *and* enforced. Thus, if a combination of two or even three procedures in this book best serves a facility, by all means adapt that combination.

Before going any further, let us define what "procedures" are. Procedures are definitive, simply written and illustrated instructions, enabling the employee to perform specific task(s) step by step. The proper sequence of steps is called the *workflow* and eventually results in finished operation, service or product.

To design, format and write functional, effective standard procedures you:

- MUST know what is the nature of the task(s); what is the most expeditious way of performing it, and; what are the *precise* steps involved in the task(s)?
- MUST know the *proper* sequence or workflow of the task(s).
- MUST be able to communicate with the employee who has to implement the procedures, and with the middle and high management who have to enforce the procedures standards.
- MUST measure each written procedure against the major and perhaps the only criterion: IS IT FUNCTIONAL?

To effect the above philosophy and produce a well-organized end product, the following steps are essential:

- Study at first hand how the experienced employee performs his/her task(s). Observe closely what the employee does, and write down every detail.
- Write up your notes, type them and show this rough draft to the employee who is performing the specific task(s), and to the employee's supervisor. Often you will find the supervisor is not really aware of what his subordinate is doing. It might also happen that the employee will deny processing paper work a certain way, or assembling or producing a product the way you stated it. In the latter case, ask the employee to "correct" your written procedures and write down how the work is "really" processed. The end result is usually only minor changes in your written procedures. What is important, though, is employee involvement; an essential variable for successful acceptance of standard procedures.

- When both the employee (clerk, operator, assembler, technician) and the supervisor agree that the procedures you wrote are correct and exactly as they want the work to be processed, then and only then should the final copy of the procedures be given to the typist.
- After proofreading carefully the final copy, add the table of contents. If the procedures involve words that are known only in that particular field, such as data processing, electronics, medical, food service, add a page of "definition of terms," so that new employees can look up "peculiar" words and phrases.
- As a final but very important step before the procedures are duplicated and distributed, get the written approval of the supervisor and (if it is the company policy) the department head or the division head. This ensures you against any possible future misunderstanding.

The standards within this book, however, are not presented to constrain creativity or to inhibit any technical/administrative procedures writer from designing a more suitable procedures format. These standards are primarily guidelines to be used as reference.

Bear in mind, however, that the best standard procedures are only as effective as the extent they are established and enforced throughout the department and company.

Chapter

6

Playscript Type Procedures

This format is one of the most popular type of procedures for manual functions because the clerk, keypunch/keyprocessing operator, assembly worker or laboratory technician can easily follow each step to perform definitive task(s).

The pages in playscript procedures are divided into two columns. The left column is used to list each task and each form to be used with the task by name and number; the right column provides a step-by-step functional description of each action.

In a procedure instructing the personnel clerk how to process changes in the employee's status, for example, in left column, under "Tasks," would state the name of the form: "PERSONNEL CHANGE NOTICE," and the number of the form: # 5879. The right column, under "Action," would say: (Step) 1. Sort PCNs by the type of change–transfer, change in salary, name change, promotion, leave of absence, termination, etc. (Step) 2. Arrange PCNs in each batch in alphabetical order. (Step) 3. Get a tickler card. (Step) 4. Write each employee's name and ID number, and the PCN's number on the tickler card. Also, today's date . . . .

There are two advantages and two disadvantages of the playscript type procedures.

ADVANTAGES

1. If there are exceptions to the stated workflow within the procedures, not only the exceptions but the reason for them can be stated without interrupting the instructions.
2. Under "NOTE" you can describe what another employee in, perhaps, another department is doing with the product, data, document, item, and so forth, before it is returned to the first employee for further or final processing. This makes for better understanding of the full cycle of processing, which in turn makes for more knowledgeable and responsible employees.

DISADVANTAGES

1. It might be difficult for some new or transferred employees to read the text, and at the same time, perform the required task(s).
2. It tends to be verbose, unless the technical/administrative procedures writer is extremely careful, even miserly, with words.

SAMPLE APPLICATION

REVOLVING CREDIT PROCESSING

DATA PROCESSING AND SYSTEMS STANDARD OPERATING PROCEDURES	DIV. Lockbox SEC.
	PROCEDURE
	PAGE
TITLE REVOLVING CREDIT PROCESSING	ISSUE DATE June 1974

32-711 (2-71)

TASK: Processing the Directly Mailed in Revolving Credit Payments, Address Changes and Correspondence

CONTROL FORMS used in this processing:

Batch Total Ticket #70-736

Posting Date Control Ticket #70-737

Transmittal Entry Letter #06-001

NOTE:

If a check comes in "Paid in Full", accept it. Make a note of it on the Revolving Credit Coupon and sort the Coupon in the "messages batch".

ACTION

Each morning receive mail.

Take the envelopes which have been opened by machine in the Lockbox Mail Room, and extract their contents.

Sort items into 2 groups: with messages/notes, and without messages/notes. -- The Revolving Credit Coupons with messages are always the last group/batch.

Check the following items:

a. Amount

If the sum on the check is larger or smaller than the "DUE" printed on the accompanying Revolving Credit Coupon, cross-out the "DUE" amount on the Coupon and write in the figure shown on check.

b. Signature

If there is no signature on the check, return check (and Coupon) to the customer with a self-addressed, stamped envelope.

c. ON-US and Other Checks

The checks sent in payment with the Revolving Credit Coupons have to be payable to "Crocker National Bank", or properly endorsed. This applies whether the check is ON-US (Crocker Bank check) or other banks.

**Through the courtesy of Crocker Bank

TASK: Processing the Directly Mailed in Revolving Credit Payments, Address Changes and Correspondence

ACTION

d. Checks Without Revolving Credit Coupons or Account Numbers

Checks received without an attached Coupon or account no., are returned to the customer with a "Request for Account Number". Enclose a self-addressed, stamped envelope.

5. Perform a "proof balance" by

a. Add up all checks on the adding machine (calculator), using double tape.

b. Add up the amounts on the coupons using the adding machine and double tape.

c. Match all batches of check totals to the batches of Coupon totals.

6. Route all batches of checks with a Transmittal Entry Letter, #06-001 to the Encoding Operator in the Equipment Room.

NOTE: The Encoding Operator processes the checks on the encoding machine. This includes MICR (Magnetic Ink Character Recognition) encoding, listing the amounts on an adding machine using a double tape, and noting on the bottom of the tape the date, batch number, and the operator's initials.

The double tapes are put on top of the check batches and routed back to the Lockbox Revolving Credit Control Clerk.

7. Check that the tape totals from the Encoding Operator and your tape total of the batches of checks agree.

8. File duplicate tape listing of both the checks and the Coupons in the appropriate Lockbox files.

**Through the courtesy of Crocker Bank

DATA PROCESSING AND SYSTEMS STANDARD OPERATING PROCEDURES	DIV. Lockbox SEC. PROCEDURE PAGE
TITLE REVOLVING CREDIT PROCESSING	ISSUE DATE June 1974

32-711 (2-71)

TASK: Processing the Directly Mailed in Revolving Credit Payments, Address Changes and Correspondence

NOTE:

The Recordak Operator processes the encoded checks through the Recordak machine. This machine endorses the checks by date and batch number, and transmits their image onto a microfilm.

The Recordak Operator processes the Coupons likewise through the Recordak machine. It dates the Coupons and transmits their image onto a microfilm.

The microfilms, containing images of the daily batches of checks and Coupons, are filed in the microfilm library of the Lockbox Division.

The checks, together with the attached Batch Total Ticket, the Transmittal Entry Letter, and the original tape listing are forwarded by the Recordak Operator to the Computer Operations on the 1st floor.

The Coupons, together with the original (adding machine) tape listing, and a Transmittal Entry Letter are forwarded to the Revolving Credit Control Clerk.

ACTION

9. Fill out a Batch Total Ticket, #70-736, for each batch of checks.

10. Fill out a Batch Total Ticket, #70-736, for each batch of Coupons.

11. Route both checks and statements with their respective Batch Total Tickets and a Transmittal Entry Letter, #06-001 to the Recordak Operator. Also the original adding machine tapes.

12. Fill out a Posting Date Control Ticket, #70-737, for the Coupons based on all the Batch Total Tickets. Put it aside until later, when work will be returned from the Recordak Operator.

13. Receive the Recordak processed Coupons together with the Batch Total Ticket, and the Transmittal Entry Letter.

14. Wrap the filled out Posting Date Control Ticket, #70-737, around the the batches of Coupons, and deliver it every afternoon to the Bank Accounting & Control Department on the 2nd floor.

15. Any correspondence is routed to the Correspondence Clerk; while address changes, if any, are given to the Keyprocessing Operator. Before forwarding them, however, make xerox copies, which are then routed to Branch Offices identified by account numbers.

** Through the courtesy of Crocker Bank

Chapter
7

Flowchart Type Procedures

Because of its high visual effect, this format is well suited for processing purchase orders, orienting new personnel in assembly, expeditor, factory and shop type of work, processing new loan applications, student registration, learning basics of electronics, carpentry, and so forth.

In writing these procedures, brevity is not a desirable attribute; it is an absolute necessity. Every word that goes into the oversized block diagram (see Figure 7-1 for definition of symbols used in the flowchart) has to be weighed against these criteria:

- Does the word move the action of the procedure ahead?
- Does the word tell the reader in the clearest possible term what he/she is supposed to do?

ADVANTAGES

1. It is easy to read and learn the procedures because there is no redundancy in the instruction, and because the tasks to be performed are easily traceable.
2. It is easy to reference any part of the procedures.

DISADVANTAGES

1. It is much more difficult to write than playscript or text type of procedures.
2. It takes much more time and effort to draw and write this format than either of the above mentioned procedures.

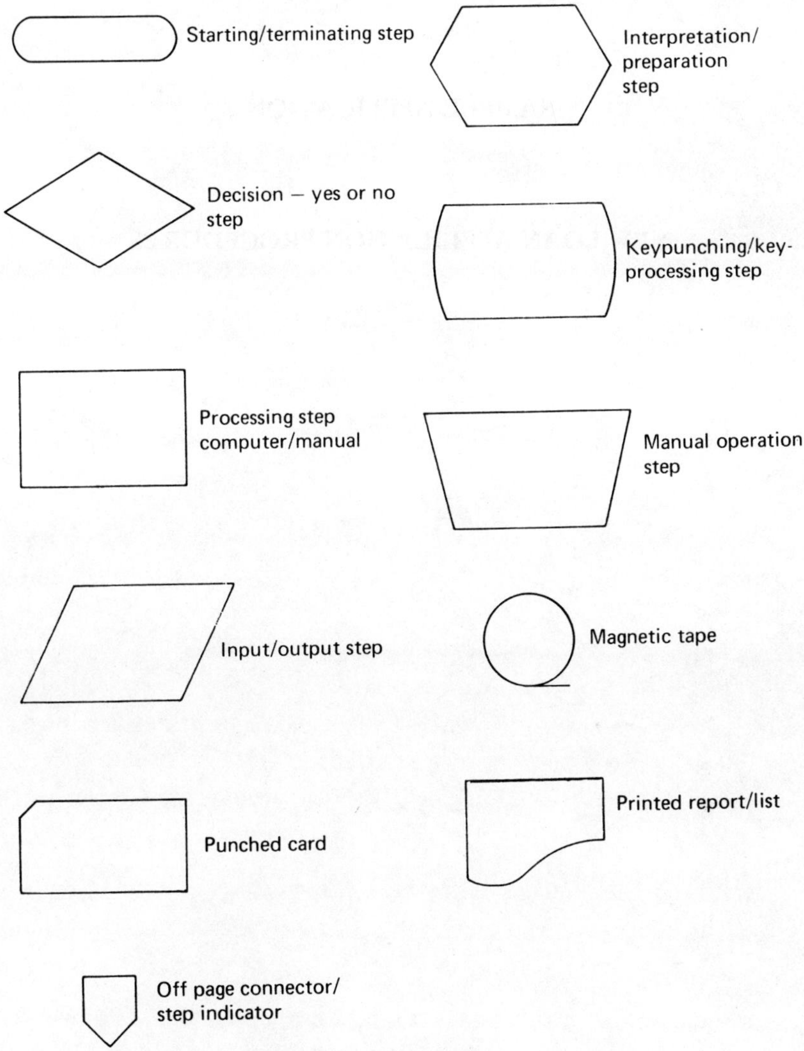

Fig. 7-1 Standard Flowchart Symbols

SAMPLE APPLICATION

NEW LOAN APPLICATION PROCEDURES

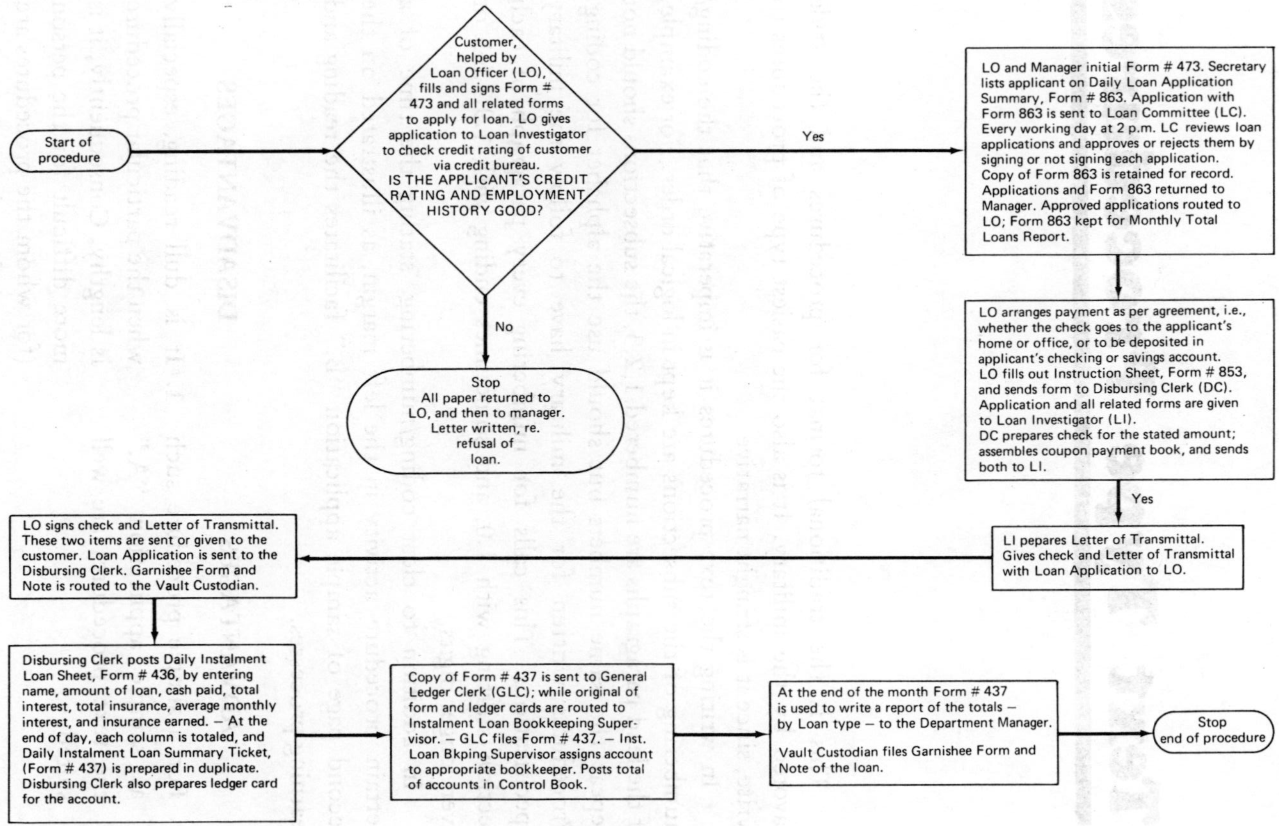
Start of procedure
Customer, helped by Loan Officer (LO), fills and signs Form # 473 and all related forms to apply for loan. LO gives application to Loan Investigator to check credit rating of customer via credit bureau. IS THE APPLICANT'S CREDIT RATING AND EMPLOYMENT HISTORY GOOD?
No
Stop All paper returned to LO, and then to manager. Letter written, re. refusal of loan.
Yes
LO and Manager initial Form # 473. Secretary lists applicant on Daily Loan Application Summary, Form # 863. Application with Form 863 is sent to Loan Committee (LC). Every working day at 2 p.m. LC reviews loan applications and approves or rejects them by signing or not signing each application. Copy of Form 863 is retained for record. Applications and Form 863 returned to Manager. Approved applications routed to LO; Form 863 kept for Monthly Total Loans Report.
LO arranges payment as per agreement, i.e., whether the check goes to the applicant's home or office, or to be deposited in applicant's checking or savings account. LO fills out Instruction Sheet, Form # 853, and sends form to Disbursing Clerk (DC). Application and all related forms are given to Loan Investigator (LI). DC prepares check for the stated amount; assembles coupon payment book, and sends both to LI.
Yes
LI pepares Letter of Transmittal. Gives check and Letter of Transmittal with Loan Application to LO.
LO signs check and Letter of Transmittal. These two items are sent or given to the customer. Loan Application is sent to the Disbursing Clerk. Garnishee Form and Note is routed to the Vault Custodian.
Disbursing Clerk posts Daily Instalment Loan Sheet, Form # 436, by entering name, amount of loan, cash paid, total interest, total insurance, average monthly interest, and insurance earned. – At the end of day, each column is totaled, and Daily Instalment Loan Summary Ticket, (Form # 437) is prepared in duplicate. Disbursing Clerk also prepares ledger card for the account.
Copy of Form # 437 is sent to General Ledger Clerk (GLC); while original of form and ledger cards are routed to Instalment Loan Bookkeeping Supervisor. – GLC files Form # 437. – Inst. Loan Bkping Supervisor assigns account to appropriate bookkeeper. Posts total of accounts in Control Book.
At the end of the month Form # 437 is used to write a report of the totals – by Loan type – to the Department Manager.
Vault Custodian files Garnishee Form and Note of the loan.
Stop end of procedure

Chapter
8

Text Type Procedures

This is the traditional format for procedures and the style favored by the military. It is also the easiest type of procedures to write, since it is straight narrative.

In writing the text procedures it is imperative that the coding/numbering of the subsections are kept in logical order. For example, if the lead paragraphs are numbered 1,2,3, the subsections should not repeat the same numbers but should use the alphabet for coding. Procedures written for the military have to follow the military specifications. This calls for numbering every paragraph of each section starting with 1.0, and sometimes ending with 1.9.7.6.5 or even longer digits.

In addition to clear coding/numbering, stating the time of a certain procedure activity in the left margin, as illustrated on the second page of sample application "B," facilitates the reading and learning processes.

ADVANTAGES	DISADVANTAGES
1. For a simple procedure such as sample application "A," the text procedures are well suited.	1. It is dull reading, especially when the particular procedure is lengthy. Consequently, it is more difficult for the person (for whom the procedures are intended) to learn the task(s).

ADVANTAGES	DISADVANTAGES
2. This format does not require special ability, and consequently beginning technical/administrative procedures writers can do the job.	2. Text type procedures are difficult to use as reference, and because of this drawback the learning period of the clerk, operator, etc., is longer. This, in turn, affects the cost of the project.

SAMPLE APPLICATION A

PERSONNEL FILE CHANGE OF NAME PROCEDURES

DATA PROCESSING AND SYSTEMS STANDARD OPERATING PROCEDURES	DIV. SEC.
	PROCEDURE
	PAGE
TITLE	
PERSONNEL FILE CHANGE OF NAME PROCEDURES	ISSUE DATE May 1974

32-711 (2-71)

Only the change of names requiring a change of employee number are in this group. Other name changes are placed in batch #3.

START OF PROCEDURE

1. Arrange Personnel Change Notices (PCNs) and the xerox copies in alphabetical order according to the prior name.

2. Make a tickler card for the prior name.

3. Arrange the PCNs in alphabetical order according to the current name.

4. Make a tickler card for the current name.

5. Assign an employee number. Write the number and the current name on the tickler card.

6. Write the old <u>and</u> new employee numbers on the PCN's comments section.

7. Put the old tickler in the file. Pull the old alpha card and stamp it "NAME CHANGED TO", and write the current name.

8. Place the current name tickler card in the file.

9. Pull and mark changes on the Payroll Work Cards (Form #5-1).

10. Attach the old alpha card to the appropriate PCN. Give the PCNs and the Work Cards to the keypunch operator to complete the personnel cycle.

11. Mark the Employee Master Card with the current name and the newly assigned employee number.

END OF PROCEDURE

Through the courtesy of Crocker Bank

SAMPLE APPLICATION B

GENERAL LEDGER SYSTEM

CONTROL PROCEDURES

3-5 GENERAL LEDGER SYSTEM CONTROL PROCEDURES

The General Ledger System provides financial management information to the Controller's Department of the Bank. At present, the system contains only income and expense (I & E) and budget data accumulated from branch entries and consolidated for total bank on a daily basis. Monthly, I & E and budget figures are also printed by cost center for each branch, region, administrative department, etc. Asset and liability entries now prepared and manually posted at the branch level will eventually be included in this automated system.

The existing front-end processing -- MICR capture (Job Name GL001), punch paper tape conversion and card capture (GL002), and the subsequent merge of these elements and interfacing applications (GL003) -- remains unchanged and, therefore, are excluded from these procedures. For details on the MICR capture, refer to Section 3-4 of this manual. To identify the new system from the old, new job names have been devised and are explained in Section 3-5-1 Job Names.

This procedure outlines the work flow of additional input from the Controller's Department to Balance & Reconcilement Section at SFDPC only, preliminary verification of a batch balance report, and fundamental control of the system process to ensure accuracy and proper completion of all jobs before reports are distributed to the Controller's Department.

3-5-1 Job Names

Job names for the General Ledger System are coded as follows:

1. GL = application initials
2. For the accounting module, no other code is designated, just frequency and sequence number. For example, GLDLY001, where:
   a. GL = application initials
   b. DLY = daily (frequency)
   c. 001 = sequence number
3. For the budget module, a B is designated in addition to frequency and sequence number. For example, GLBONR01, where:
   a. GL = application initials
   b. B = budget
   c. ONR = on request (frequency)
   d. 01 = sequence number

Through the courtesy of Crocker Bank

4. For the support module, an S is designated in addition to frequency and sequence number. For example, GLSYLY01, where:

   a. GL = application initials

   b. S = support

   c. YLY = yearly (frequency)

   d. 01 = sequence number

   NOTE: Frequency abbreviations are:

DLY = daily	QTR = quarterly
WKY = weekly	YLY = yearly
MON = monthly	ONR = on request

3-5-2 Daily Cycle

After all income and expense (I & E) capture programs have been run (GL001 through GL003) and the LA TRANMERGE has been transmitted, Balance & Reconcilement begins processing the new General Ledger System. The following schedule approximates the daily procedure:

By 6:00 p.m.	Balance & Reconcilement Section receives Controller's input and prepares work for key processing.
By 12:00 midnight	Keyed input is completed by Key Processing Section and returned to Balance & Reconcilement. Balance & Reconcilement submits GLDLY001 job to Production Control Section.
By 4:00 a.m.	GL003 and OSRECEIVE of LA TRANMERGE jobs are normally completed by Computer Processing Section. Balance & Reconcilement receives output from Production Control and verifies GLDLY001 output. Then this section submits GLDLY002 job to Production Control.
By 6:00 a.m.	Balance & Reconcilement balances GLR010 Transaction Batch Balance Listing and submits GLDLY003 - GLDLY007 and any on-request jobs to Production Control.
By 8:00 a.m.	Output reports are forwarded to Distribution Section for decollating and distribution. COM tapes are sent to Micrographics Section for microfiche processing.
By 9:30 a.m.	All daily reports are received in Controller's Department.

Through the courtesy of Crocker Bank

DATA PROCESSING AND SYSTEMS STANDARD OPERATING PROCEDURES	DIV. SEC.
	PROCEDURE
	PAGE
TITLE	
GENERAL LEDGER SYSTEM CONTROL PROCEDURES	ISSUE DATE May 1974

3-5-2-1 Disposition of Daily Documents.

1. Place the following in a box and route to Controller's Department:

   a. All hardcopy reports (1-part paper)

   b. Original of Input Control Totals Sheet

   c. All MICR tickets

   d. Punched paper tape

2. Forward all reports printed on 3-part paper to Utility Section for decollating and final distribution.

3. Staple together all JSFs for the current day's processing and file for 6 months reference.

4. Forward any punchout from GLDLY001 to Unit Record Section for interpreting before forwarding cards to Controller's Department.

   See 3-5-2-13 for Interpreting Punchout instructions.

5. Include entry letters (Part 2 copies) prepared for rejects in day's I & E blotter.

6. Retain the following in department files:

   a. Input documents for 7 days.

   b. GLDLY001 keypunched cards for 3 days.

   c. Xerox copies of balancing control documents in file folder marked General Ledger Transmittals for Balancing.

   d. SYSOUTs for 7 days.

Through the courtesy of Crocker Bank

Chapter
9

Text-Flowchart Type Procedures

This format—a combination of the text and flowchart types of procedures—is not as popular as the playscript type, even though it has a lot of plus features. It is visual and it gives written instructions that complement and clarify the flowchart.

The pages in the text-flowchart procedures are divided into two parts, the same as the playscript procedures. The only difference being that in this format the left side of the page is used for a work flowchart that parallels the text on the right side. To facilitate the learning process each step (next to the appropriate symbol) is numbered. The numbers correspond to the numericals in the text side of the page.

The text-flowchart procedures can be made as simple or as sophisticated as the type of work and level of personnel warrant it. In the simple format the work flowchart keeps in step with the written instructions. In the more sophisticated format the flowchart indicates not only the immediate action but also the outputs resulting from a certain action and the destination of these outputs. Moreover, via the "off-page" symbol the future steps are specified. Often the simple and the more complex formats are combined, i.e., the procedures start with the former and by the second page or so evolve into the latter.

ADVANTAGES

1. The personnel involved in the particular processing—

DISADVANTAGES

1. This format is more work for the technical/administrative

ADVANTAGES

whether purchase orders, installment loan payments, payroll, or whatever—get the "whole picture" via the work flowchart that winds through the left side of all pages describing the procedures.

2. The work flowchart, illustrating the procedures, breaks the monotony of the text and, consequently, it is easier for the employee to learn the particular procedures.

DISADVANTAGES

procedures writer than either the playscript or the flowchart procedures.

2. For personnel who are not familiar with the flowchart symbols, it might be a little difficult at first to read the flowchart.

SAMPLE APPLICATION

AUTO FINANCE CONTROL PROCEDURES

DATA PROCESSING AND SYSTEMS STANDARD OPERATING PROCEDURES	DIV. SEC.
	PROCEDURE
	PAGE
TITLE AUTO FINANCE CONTROL PROCEDURES	ISSUE DATE 5/1/74

32-711 (2-71)

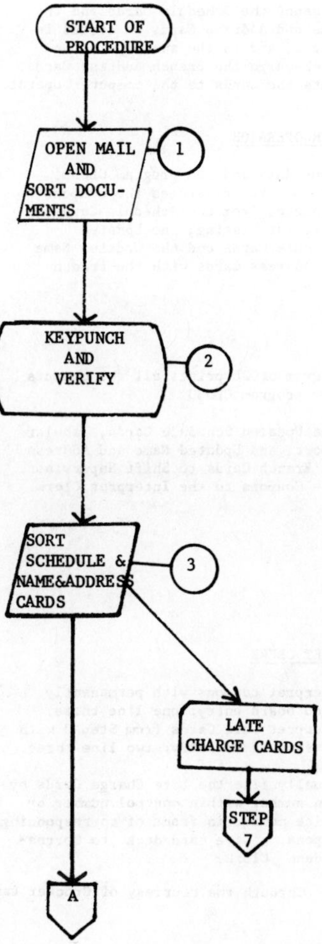

MAIL CLERK

1. Entry forms 70 &72 with documents received daily from the Branch Offices by 8:30 a.m. in a large yellow envelope. Open mail, stamp date received, and sort documents into Branch Office number sequence.Route batches to Keypunch section.

KEYPUNCH OPERATOR

2. Punch and verify schedule, name, and address, and late charge cards. Forward completed cards to Control Clerk.

CONTROL CLERK

3. Sort Schedule Cards thusly:

Card sequence no.	columns	29-28	numerical
Loan number	"	38-24	"
Account number	"	43-39	"
Office number	"	32-30	

Sort Name and Address Cards thusly:

Card sequence code	columns	80	alphabetical
Loan number	"	9-5	numerical
Account number	"	14-10	"
Office number	"	3-1	"

Give the cards to the data processing clerk, and entry forms 70 and 72 to the Shift Supervisor.

Through the courtesy of Crocker Bank

DATA PROCESSING AND SYSTEMS
STANDARD OPERATING PROCEDURES

DIV.	SEC.
PROCEDURE	
PAGE	
ISSUE DATE	5/1/74

TITLE

AUTO FINANCE CONTROL PROCEDURES

32-711 (2-71)

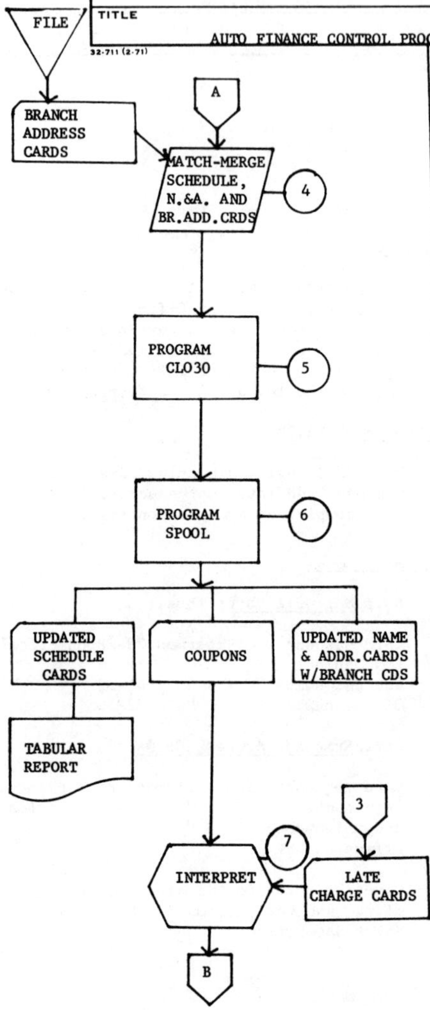

DATA PROCESSING CLERK

4. Sequence the Schedule Cards and the Name and Address Cards as sorted in step 3, and in the same operation match-merge the Branch Address Cards. Route the cards to the computer operator.

COMPUTER OPERATOR

5. Input data and run program CL030. This program generates the Coupons (produced from the Schedule Cards), a tabular listing, the Updated Schedule Cards and the Updated Name and Address Cards with the Branch Cards.

6. Program SPOOL prints all the outputs from program CL030.

   Give Updated Schedule Cards, Tabular Report, and Updated Name and Address and Branch Cards to Shift Supervisor. Give Coupons to the Interpret Clerk.

INTERPRET CLERK

7. Interpret coupons with permanently wired board entry: one line three. Interpret Late Cards from Step 3 with the same board entry: two line three.

   Manually file the Late Charge Cards by loan number within control number by office number in front of corresponding coupons. Give card deck to Correspondent Clerk.

Through the courtesy of Crocker Bank

DATA PROCESSING AND SYSTEMS STANDARD OPERATING PROCEDURES	DIV. SEC.
	PROCEDURE
	PAGE
TITLE AUTO FINANCE CONTROL PROCEDURES	ISSUE DATE 5/1/74

32-711 (2-71)

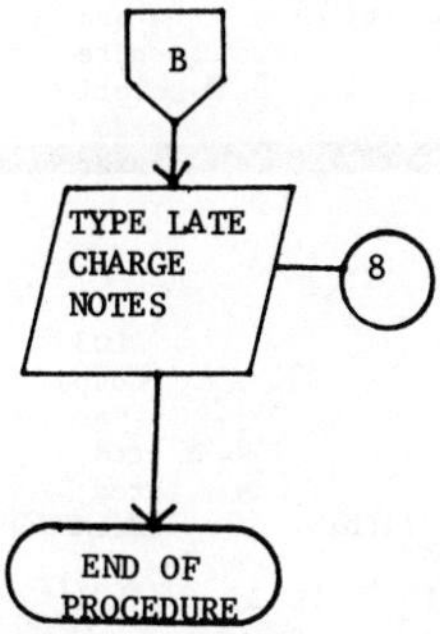

CORRESPONDENT CLERK

8. Type Late Charge Notes in triplicate. Give whole package to Shift Supervisor. He checks this and all the previous documents against entry forms 70 and 72; keeps one copy of the Late Charge Notes, and sends the other two copies together with all the updated material to the Branch Offices.

Through the courtesy of Crocker Bank

Chapter
10

Picture Type Procedures

This terse, clearly visible procedures format is favored by engineering and construction companies, though it is equally well suited for most other fields such as banking, data processing, manufacturing, marketing and the like.

Since these procedures rely heavily on drawings, a lot of preliminary work goes into them before the final diagrams are drawn on large scale and then reduced to effect the summary yet lucid form (usually) on one page. For legibility, the economically worded instructions are typed on the standard (8½ x 11) page after the drawings have been reduced.

The following four sample applications present variations of the picture procedures. The "note" added to the fourth sample is for better understanding of what is involved in the particular procedure, and shows the flexibility of this format.

ADVANTAGES

1. Its all around economy—paper, text and time of comprehension—is undisputable.

DISADVANTAGES

1. Unless the administrative/technial procedures writer is very careful, this format can become "too busy," and it can confuse rather than guide the employee in performing the task(s).

ADVANTAGES	DISADVANTAGES
2. It is excellent as a reference and memory-jogger.	2. Because the drawings and the inked-in notations are reduced (sometimes quite drastically to have a complete procedure fit one page), this format is not for people with limited eyesight.

SAMPLE APPLICATION

VARIOUS TYPES OF ADMINISTRATIVE INPUT

ADMIN. HQ. INPUT TO SFDPC
CONTROLLER'S DEPARTMENT

CERTIFICATE OF DEPOSIT
SYSTEM MAINTENANCE WORK

BATCH & SEND CERTIFICATE OF DEPOSIT SYSTEM MAINTENANCE WORK

Form	Purpose
CERTIFICATE OF DEPOSIT SET UP FORM (UNNUMBERED FORM)	To request computer-run reports in alpha name order
HOLIDAY TABLE SET UP FORM (UNNUMBERED FORM)	To add or delete holidays in the CD system
GENERAL LEDGER SET UP FORM (UNNUMBERED FORM)	To add or delete types of CD's
ACCOUNT BALANCE MASTER FILE MAINTENANCE SET UP FORM (UNNUMBERED FORM)	To show CD types by office; one card for each type offered by each office
BRANCH TABLE SET UP FORM (UNNUMBERED FORM)	To set up a new branch office in the CD system

Batch work as shown, band it securely, and place it in a plain envelope addressed to Certificate of Deposit Control, SFDPC; place the envelope in the interoffice mail.

Through the courtesy of Crocker Bank

ADMIN. HQ. INPUT TO SFDPC
DATA PROCESSING + SYSTEMS ADMIN.
FINANCIAL ANALYSIS SECTION

PROFIT ANALYSIS

1 BATCH & SEND WORK FOR KEYPROCESSING

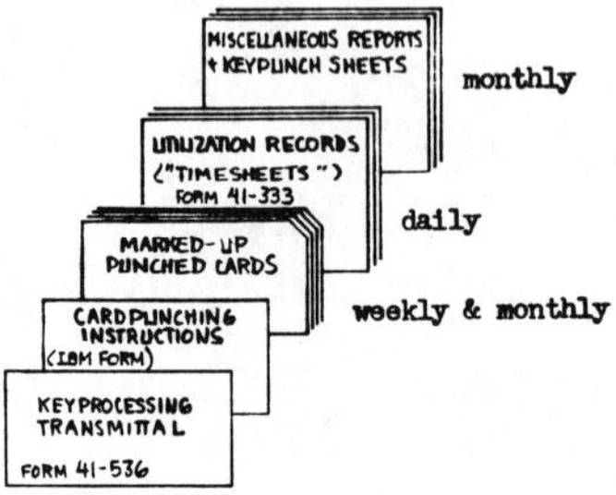

Batch completed work as shown and carry it to Keyprocessing, SFDPC.

2 BATCH & SEND UNIT RECORD WORK

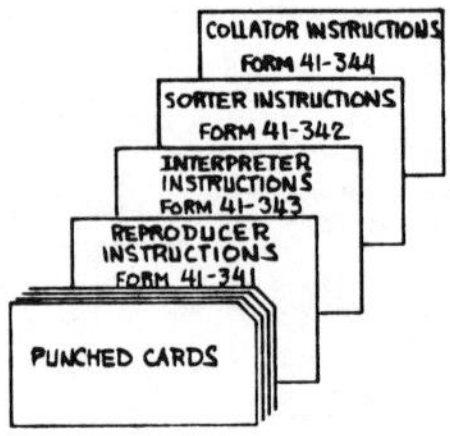

Batch completed work as shown and carry it to Unit Record, SFDPC.

3 BATCH & SEND WORK FOR COMPUTER PROCESSING

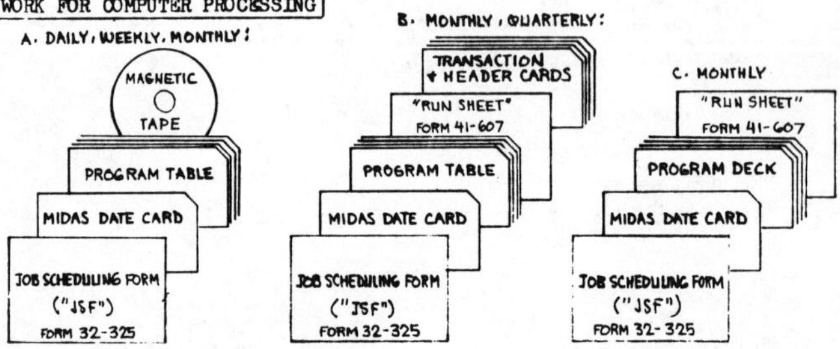

Time & Cost Reports

Centrex & Supply Reports

Batch work as shown and carry to Production Control & Information (PCI), SFDPC.

Through the courtesy of Crocker Bank

ADMIN. HQ. INPUT TO SFDPC
INSTALMENT CREDIT ADMIN., RCCD
NON-MONETARY SECTION

MASTER CHARGE
DAILY OCR WORK

BATCH & SEND DAILY OCR WORK

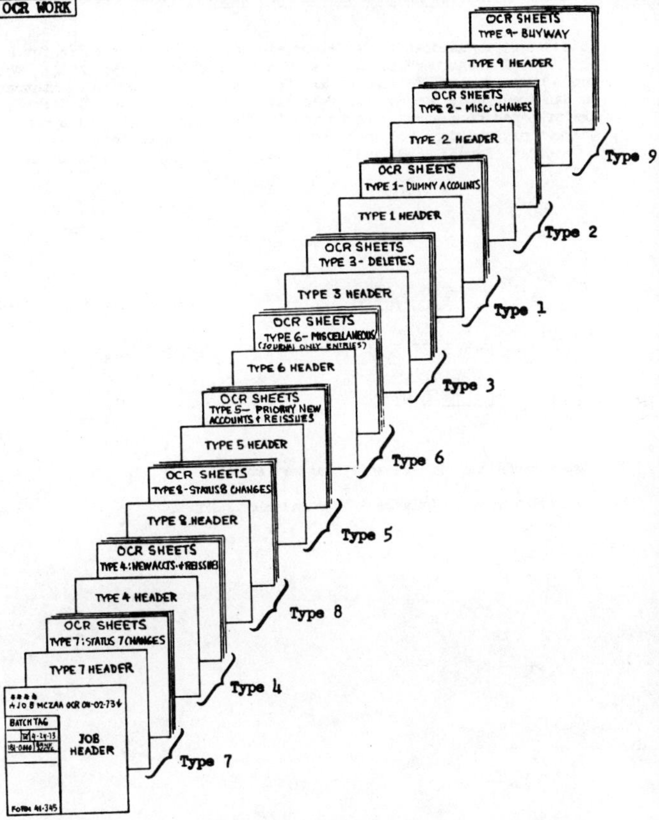

1. Set up daily OCR work as shown and complete a Batch Tag (Form 41-345) as shown.
2. Band the work and hand-carry it to Master Charge Control, Bank Accounting, SFDPC.

Through the courtesy of Crocker Bank

ADMIN. HQ. INPUT TO SFDPC
DATA PROCESSING SYSTEMS ADMIN.
FINANCIAL ANALYSIS SECTION

INCOME + EXPENSE
MONTHLY CENTREX ALLOCATIONS

MONTHLY, PREPARE & SEND I&E CENTREX ALLOCATION CARDS

Note: Centrex allocation cards contain I&E debits and credits for 4 accounts: 22356 Telephone Service Charges; 22450 Message Units; 22765 Toll Charges; and 22478 Federal Telephone Tax; for cost centers served by Centrex equipment in San Francisco, Sacramento, Palo Alto, Los Angeles, and El Segundo. The cards are prepared from summary cards received monthly from the telephone companies in these areas and from keypunch sheets prepared by operators showing "Other Charges & Credits" (Form 31-520) and "Telephone Relocation" charges (unnumbered keypunch sheet).

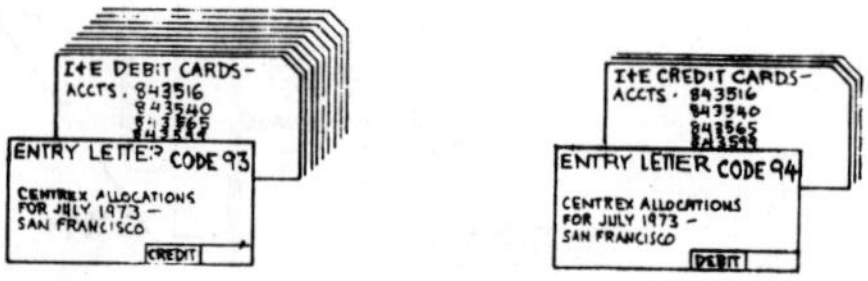

1. Batch cards and prepare entry letters as shown for the totals.
2. Carry the work to Balance & Reconcilement, SFDPC.

Through the courtesy of Crocker Bank

Chapter
11

Text-Picture Type Procedures

This format is a merging of text and picture types of procedures. It is used for processes in which the terse instructions of the picture procedures are not enough for the employee to learn a particular operation.

In this format too, however, the diagrams are drawn on large scale and subsequently reduced. The concisely written instructions are typed on the standard page after the drawings have been reduced.

A variation of the reduced drawings and inked-in notations method is presented in Sample Application B. In this procedure the explanation of various forms used, and other clarifications are typed into the already reduced diagrams. Few people favor this method, since the inked-in notations—reduced along with the schematics—can be difficult to read.

ADVANTAGES

1. For economy-minded management, this format should appeal almost as much as the picture procedures.
2. It is very good as reference.

DISADVANTAGES

1. Drawing and writing this format requires more time and a higher degree of capability than some other types of procedures.
2. This format too can be a bit hard on the reader/learner's eyes.

SAMPLE APPLICATION A

VARIOUS TYPES OF ADMINISTRATIVE INPUT

ADMIN. HQ. INPUT TO SFDPC
CONTROLLER'S DEPARTMENT

AUTO FINANCE DISCOUNT LOAN

1 DAILY, BATCH & SEND CONTRACTS AND MISCELLANEOUS CREDITS

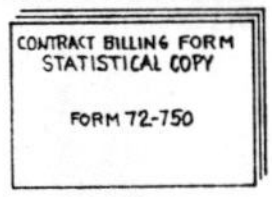

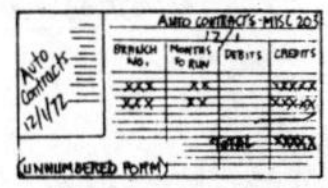

A. Daily, from I&E tickets processed by the DPC and sent to the Controller's Department, select all I&E credit tickets to Acct. 859082, Unearned Discount-Auto Finance Loans, sort the tickets by date, and check them off against the I&E Transaction Journal to make sure each was correctly posted. Also check the reverse of each ticket to make sure that months-to-run figures are marked (if not, telephone the branch, obtain the figures, and enter them on the ticket). Process the I&E debit tickets as shown in Frame 3 below.

B. Check the credit tickets against the day's contracts received from Installment Credit Administration. For all amounts shown on the tickets but not shown on the contracts, make miscellaneous credit entries as shown on the form illustrated at upper right, and total them on the form.

C. Run a tape listing of the contracts by branch office, with branch numbers non-added, and staple the tape to the miscellaneous credit form as shown. The total of the tape plus the total of the miscellaneous credit entries must equal the total of all credit tickets. If out of balance, re-check work and make corrections.

D. Bundle the contracts and the miscellaneous credit form together and place them in a plain envelope addressed to Auto Discount Loan Control, Bank Accounting Section, SFDPC, and place the envelope in the interoffice mail for transmission to the DPC.

<u>Note</u>: I&E credit tickets are not sent to the DPC.

2 MONTHLY, SEND "LATE CONTRACTS"

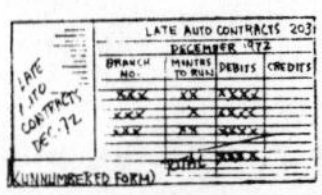

A. When an I&E credit ticket is received without a matching contract, the amount of the ticket is entered as a miscellaneous credit (see 1, B, above). When the matching contract is received, it is held until the end of the month.

B. Enter amounts of late contracts on a Late Contract sheet as shown (as debits, to reverse them out of misc. credits) & tape list their totals as shown.

C. At the end of the month, bundle the late contracts and the Late Contracts sheet, place them in a plain envelope addressed to Auto Discount Loan Control, and place the envelope in the interoffice mail for transmission to the DPC.

3 MONTHLY, SEND I&E DEBITS & ADJUSTMENTS

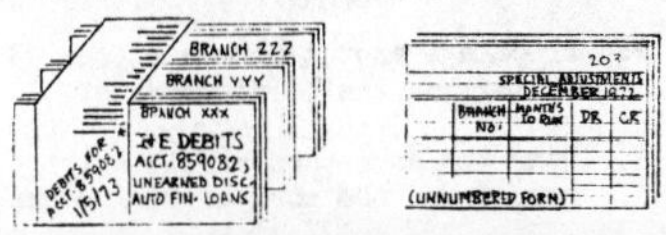

A. Hold all I&E debit tickets received until the 4th or 5th of the next month; then sort and tape list them by branch; wrap the tape around the tickets; and place them in a plain envelope addressed to Auto Discount Loan Control.

B. In the same envelope, include any Special Adjustment sheets required to correct errors made in processing the daily work shown in Frame 1 above.

C. Place the envelope in the interoffice mail for transmission to the DPC.

Through the courtesy of Crocker Bank

ADMIN. HQ. INPUT TO SFDPC
VARIOUS DEPARTMENTS (SEE INDEX) **DUE-FROM CORRESPONDENT ENTRIES**

BATCH & SEND DUE-FROM CORRESPONDENT ENTRIES

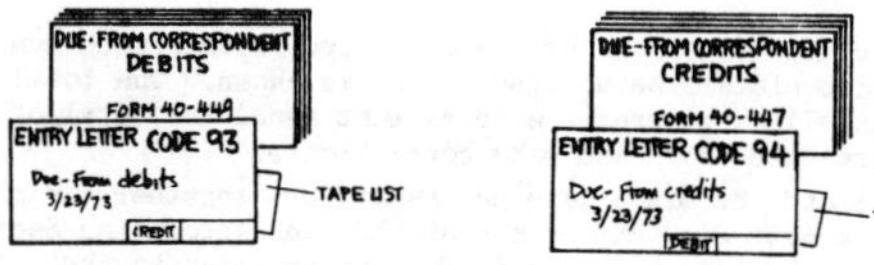

1. **Batch debit and credit entries separately as shown and tape list each, marking the tapes with the department name and number, and the date.**
2. **Prepare & amount-encode a Code 93 entry letter for the total debits, a Code 94 entry letter for the total credits.**
3. **Band the debit and credit entries separately, with the entry letters on top, and hand-carry the work to Bank Books, SFDPC.**

Through the courtesy of Crocker Bank

BATCH & SEND CHECKS RECEIVED IN PAYMENT FOR SERVICES

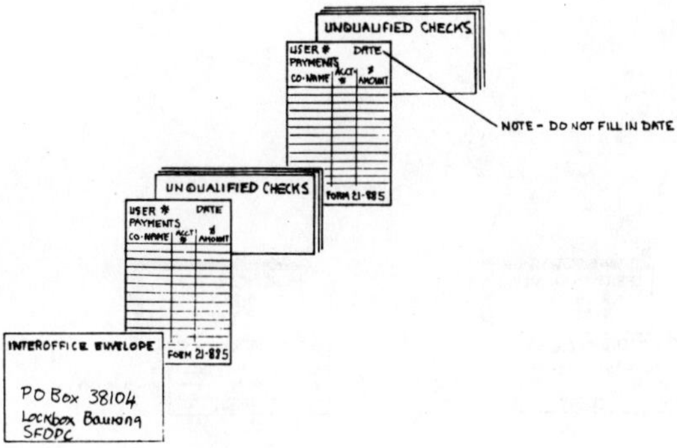

1. Sort loose checks by the user number to which they are to be credited.

2. Fill out a payment sheet for each user number with

   a. User number at the top of the form.
   b. Company name from the check
   c. Company's account number from the user file
   d. Dollar amount of the check.

3. Clip checks to the back of the payment sheets prepared for them and place them in an interoffice envelope addressed to Lockbox, SFDPC. Place the envelope in the red bag for transmission to the DPC.

Through the courtesy of Crocker Bank

BATCH & SEND GENERAL LEDGER ENTRIES

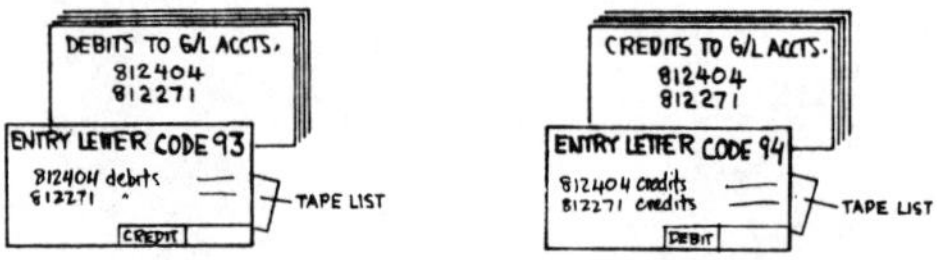

1. Batch debit and credit entries separately for each account and run tape lists. At the top of the debit tape list, run entries for Acct. 812404 and subtotal; at the bottom, run entries for Acct. 812271 and subtotal; and strike a grand total. Do the same with the credit tape. Mark both tapes "Investment #560" plus the date.

2. Prepare and amount-encode a Code 93 entry letter for the total debits, a Code 94 entry letter for the total credits.

3. Band the debit and credit entries separately, with the entry letters on top, and hand-carry the work to Input/Output, SFDPC.

Through the courtesy of Crocker Bank

SAMPLE APPLICATION B

COMPOSITE CHECKS CONTROL PROCEDURES

BATCHING

1. Arrange agency listings and single items for CC processing into batches of a reasonable size for key machine operators:

   a. An agency listing from 1 to 4 pages long is treated as a single batch and is assigned a separate batch number;

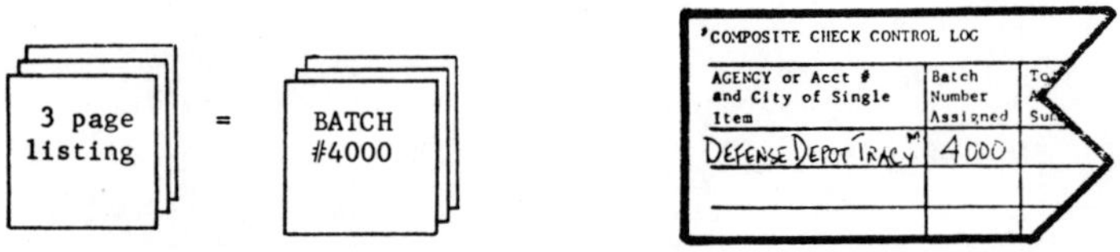

   b. An agency listing longer than 4 pages must be divided into 2 or more smaller batches no longer than 4 pages each:

      (1) Compute the deposit total for each of these new batches;

      (2) Make certain that the sum of these new batch totals equals the listing total;

      (3) Assign a batch number to each new batch unit;

   c. Agency listings with very few accounts, and several single items may be grouped together to form a batch large enough to merit key processing:

      (1) Write the account numbers and deposit amounts of each individual check on one sheet of paper and attach any short listings;

Through the courtesy of Crocker Bank

(2) Compute the total deposits for all accounts included in the batch;

(3) Make sure this batch total equals the sum of all the single item checks and the listing totals.

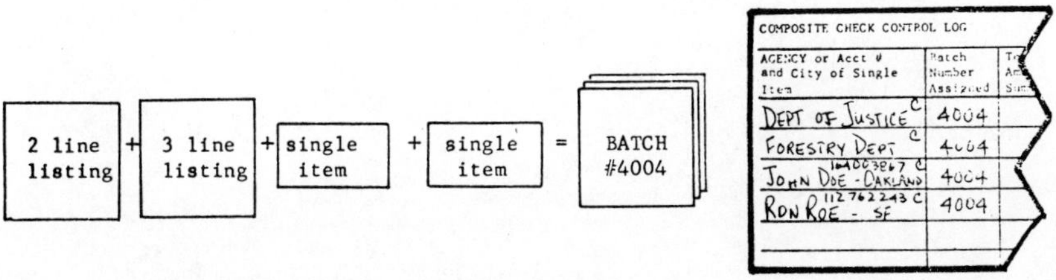

2. If there are too few single items and small listings to justify key processing, process these items manually but assign them all the same batch number in the Control Log. (See Exhibit B, "Manual Handling")

NOTE: Assign batch numbers starting with number 4000 each day running consecutively to 4999 if necessary.

Through the courtesy of Crocker Bank

Index

A

C

D

F

G

I

N

O

P

Q

R

S

U

W

Notes

NOTES

NOTES

NOTES

NOTES

NOTES

NOTES

NOTES